ADVANCED POLYMERIC MATERIALS

MATERIALS

From Macro- to Nano-Length Scales

ADVANCED POLYMERIC MATERIALS

MATERIALS

From Macro- to Nano-Length Scales

Edited by
Sabu Thomas, PhD
Nandakumar Kalarikkal, PhD
Maciej Jaroszewski, PhD
Josmin P. Jose

APPLE
ACADEMIC
PRESS

Apple Academic Press Inc. | Apple Academic Press Inc.
3333 Mistwell Crescent | 9 Spinnaker Way
Oakville, ON L6L 0A2 | Waretown, NJ 08758
Canada | USA

© 2016 by Apple Academic Press, Inc.

First issued in paperback 2021

Exclusive worldwide distribution by CRC Press, a member of Taylor & Francis Group

No claim to original U.S. Government works

ISBN 13: 978-1-77463-372-4 (pbk)
ISBN 13: 978-1-77188-096-1 (hbk)

Library and Archives Canada Cataloguing in Publication

Advanced polymeric materials : from macro- to nano-length scales / edited by Sabu Thomas, PhD, Nandakumar Kalarikkal, PhD, Maciej Jaroszewski, PhD, Josmin P. Jose.

Includes bibliographical references and index.
Issued also in electronic format.
ISBN 978-1-77188-096-1 (hardcover)
1. Polymers. 2. Polymeric composites. I. Thomas, Sabu, editor II. Jose, Josmin P., editor III. Kalarikkal, Nandakumar, editor IV. Jaroszewski, Maciej, editor

TA455.P58A38 2015   620.1'92   C2015-905413-3

CIP data on file with US Library of Congress

Apple Academic Press also publishes its books in a variety of electronic formats. Some content that appears in print may not be available in electronic format. For information about Apple Academic Press products, visit our website at **www.appleacademicpress.com** and the CRC Press website at **www.crcpress.com**

CONTENTS

List of Contributors vii

List of Abbreviations xi

Preface xv

About the Editors xvii

1. Polymer Hydrogel Dressing in Wound Management 1

Ajith James Jose, Vinu Varghese, and Sam John

2. AC Impedance and FTIR Studies on Proton Conducting
   Polymer Electrolyte Membrane Based on PVP and
   Methanesulfonic Acid 17

C. Ambika, G. Hirankumar, S. Karthickprabhu, and R. S. Daries Bella

3. Dendritic Organic Semiconductors Based on Pyrene
   and Triazine Derivatives 31

Renji R. Reghu and Juozas V. Grazulevicius

4. Rheological Characteristics of Linear Low Density
   Polyethylene – Fumed Silica Nanocomposites 61

V. Girish Chandran and Sachin Waigaonkar

5. Rational Design of Molecularly Imprinted Polymers: A Density
   Functional Theory Approach 77

Shashwati Wankar, Aakanksha Jha, and Reddithota J. Krupadam

6. Preparation of Polymer and Ferrite Nanocomposites for
   EMI Applications 99

P. Raju and S. R. Murthy

7. Macro Level Investigation on Thickness Variation of
   Rot Moulded LLDPE Product 121

P. L. Ramkumar, Dhananjay M. Kulkarni, and Sachin D. Waigaonkar

8. Molecular Structure and Property Relationship of Commercial
    Biaxially Oriented Polypropylene (BOPP) by DSC, GPC and
    NMR Spectroscopy Techniques 135

    Ravindra Kumar, Veena Bansal, S. Mondal, Nitu Singh, A. Yadav,
    G. S. Kapur, M. B. Patel, and Shashikant

9. Lead and Cadmium Ion Removal by Novel Interpenetrating
    Polymer-Ceramic Nanocomposite 143

    K. Sangeetha and E. K. Girija

10. Microwave Assisted Synthesis of Polyacrylamide Grafted
    Casein (CAS-g-PAM) as an Effective Flocculent for
    Wastewater Treatment 153

    Sweta Sinha, Gautam Sen, and Sumit Mishra

11. Polymer Assisted Synthesis of CdS Nanostructure for
    Photoelectrochemical Solar Cell Applications 173

    S. A. Vanalakar, J. H. Kim, and P. S. Patil

12. Synthesis and Characterization Chitosan-Starch Crosslinked
    Beads 205

    Virpal Singh

13. Dendrimer Polymer Brushes 219

    Wei Cui, Holger Merlitz, and Chen-Xu Wu

14. Photo-Bactericidal Polyacrylonitrile Matrix for
    Protective Apparels 227

    G. Premika, K. Balasubramanian, and Kisan M. Kodam

    Index 237

LIST OF CONTRIBUTORS

C. Ambika
C-SAR, School of Basic Engineering and Sciences, PSN College of Engineering and Technology, Melathediyoor, Tirunelveli, Tamilnadu, India; E-mail: g.hirankumar@psnresearch.ac.in

K. Balasubramanian
Department of Materials Engineering, Defence Institute of Advanced Technology, Ministry of Defence, Girinagar, Pune 411025, India; Email: meetkbs@gmail.com, balask@diat.ac.in

Veena Bansal
Indian Oil Corporation Limited, Research and Development Centre, Sector-13, Faridabad–121007, India

R. S. Daries Bella
C-SAR, School of Basic Engineering and Sciences, PSN College of Engineering and Technology, Melathediyoor, Tirunelveli, Tamilnadu, India; E-mail: g.hirankumar@psnresearch.ac.in

V. Girish Chandran
Department of Mechanical Engineering, BITS Pilani K K Birla Goa Campus, Zuarinagar Goa, India–403726; Email: p2011407@goa.bits-pilani.ac.in; sdw@goa.bits-pilani.ac.in

Wei Cui
Department of Physics and ITPA, Xiamen University, 361005 Xiamen, China; E-mail: cuiwei1008@163.com

E. K. Girija
Department of Physics, Periyar University, Salem–636 011, India; Tel.: +91 9444391733; Fax: +91 427 2345124; E-mail: girijaeaswaradas@gmail.com

Juozas V. Grazulevicius
Department of Polymer Chemistry and Technology, Kaunas University of Technology, Radvilenu pl. 19, LT-50254 Kaunas, Lithuania, Tel: +37037 300193; Fax: +37037 300152; E-mail: juozas.grazulevicius@ktu.lt

G. Hirankumar
C-SAR, School of Basic Engineering and Sciences, PSN College of Engineering and Technology, Melathediyoor, Tirunelveli, Tamilnadu, India; E-mail: g.hirankumar@psnresearch.ac.in

Maciej Jaroszewski
High Voltage Laboratory, Wroclaw University of Technology, Wybrzeze Wyspianskiego 27, 50-370 Wroclaw, Poland; E-mail: maciej.jaroszewski@pwr.wroc.pl

Aakanksha Jha
National Environmental Engineering Research Institute, Jawaharlal Nehru Marg, Nagpur 440020, India; Tel.: +91-712-2249884; Fax: +91-712-2249896; E-Mail: rj_krupadam@neeri.res.in

Sam John
Research and Postgraduate Department of Chemistry, St. Berchmans College, Changanacherry, Kerala 686 101, India

Josmin P. Jose
School of Chemical Sciences, Mahatma Gandhi University,
Kottayam, Kerala, India–686560; E-mail: josminroselite@gmail.com

Ajith James Jose
Research and Postgraduate Department of Chemistry, St. Berchmans College, Changanacherry,
Kerala 686 101, India

Nandakumar Kalarikkal
International and Inter University Centre for Nanoscience and Nanotechnology, and School of Pure
and Applied Physics, Mahatma Gandhi University, Kottayam, Kerala, India–686560;
E-mail: nkkalarikkal@mgu.ac.in

G. S. Kapur
Indian Oil Corporation Limited, Research and Development Centre, Sector-13, Faridabad-121007,
India

S. Karthickprabhu
Department of Physics, Kalasalingam University, Krishnankoil, Virudhunagar, Tamilnadu, India

J. H. Kim
Department of Materials Science and Engineering, Chonnam National University, Gwangju –
500 757, South Korea

Kisan M. Kodam
Department of Chemistry, University of Pune, Pune 411030, India

Reddithota J. Krupadam
National Environmental Engineering Research Institute, Jawaharlal Nehru Marg, Nagpur 440020,
India; Tel.: +91-712-2249884; Fax: +91-712-2249896; E-Mail: rj_krupadam@neeri.res.in

Dhananjay M. Kulkarni
BITS PILANI, K. K. Birla Goa campus, Goa–403726, India

Ravindra Kumar
Indian Oil Corporation Limited, Research and Development Centre, Sector-13, Faridabad-121007,
India; E-mail: kumarr88@indianoil.in

Holger Merlitz
Leibniz-Institute of Polymer Research, 01069 Dresden, Germany

Sumit Mishra
Department of Applied Chemistry, Birla Institute of Technology, Mesra, Ranchi – 835215,
Jharkhand, India

S. Mondal
Indian Oil Corporation Limited, Research and Development Centre, Sector-13, Faridabad-121007,
India

S. R. Murthy
Department of Physics, Osmania University, Hyderabad–500 007, India

M. B. Patel
Indian Oil Corporation Limited, Research and Development Centre, Sector-13, Faridabad-121007,
India

P. S. Patil
Thin Film Materials Laboratory, Department of Physics, Shivaji University, Kolhapur-416204, M.S., India

G. Premika
Department of Materials Engineering, Defence Institute of Advanced Technology, Ministry of Defence, Girinagar, Pune 411025, India

P. Raju
Department of Physics, Osmania University, Hyderabad–500 007, India

P. L. Ramkumar
BITS PILANI, K. K. Birla Goa campus, Goa–403726, India; Mobile: +919823256780; Email: ramkumarpl@goa.bits-pilani.ac.in, plramkumarno1@gmail.com

Renji R. Reghu
Department of Polymer Chemistry and Technology, Kaunas University of Technology, Radvilenu pl. 19, LT-50254 Kaunas, Lithuania, Tel: +37037 300193; Fax: +37037 300152; E-mail: juozas.grazulevicius@ktu.lt

K. Sangeetha
Department of Physics, Periyar University, Salem–636 011, India

Gautam Sen
Department of Applied Chemistry, Birla Institute of Technology, Mesra, Ranchi – 835215, Jharkhand, India

Shashikant
Indian Oil Corporation Limited, Research and Development Centre, Sector-13, Faridabad-121007, India

Nitu Singh
Indian Oil Corporation Limited, Research and Development Centre, Sector-13, Faridabad-121007, India

Virpal Singh
Department of Chemical Technology, Sant Longowal Institute of Engineering and Technology, Longowal, Sangrur, Punjab–148106, India, E-mail: singh_veer_pal@rediffmail.com

Sweta Sinha
Department of Applied Chemistry, Birla Institute of Technology, Mesra, Ranchi – 835215, Jharkhand, India; Tel.: +91 9801334228; E-mail: sweta.sinha2203@gmail.com

Sabu Thomas
International and Inter University Centre for Nanoscience and Nanotechnology, Mahatma Gandhi University, Priyadarshini Hills P. O., Kottayam, Kerala–686560, India; E-mail: sabuchathukulam@yahoo.co.uk, sabupolymer@yahoo.com

S. A. Vanalakar
Department of Materials Science and Engineering, Chonnam National University, Gwangju–500 757, South Korea

Vinu Varghese
Quality Control Department, Sance Laboratories Private Limited, Pala, Kottayam, Kerala 686573, India

Sachin D. Waigaonkar
Department of Mechanical Engineering, BITS Pilani K K Birla Goa Campus, Zuarinagar, Goa, India–403726; Email: p2011407@goa.bits-pilani.ac.in; sdw@goa.bits-pilani.ac.in

Shashwati Wankar
National Environmental Engineering Research Institute, Jawaharlal Nehru Marg, Nagpur 440020, India; Tel.: +91-712-2249884; Fax: +91-712-2249896; E-Mail: rj_krupadam@neeri.res.in

Chen-Xu Wu
Department of Physics and ITPA, Xiamen University, 361005 Xiamen, China

A. Yadav
Indian Oil Corporation Limited, Research and Development Centre, Sector-13, Faridabad-121007, India

LIST OF ABBREVIATIONS

ACN	acetonitrile
AFM	atomic forced microscopy
AIBN	2,2'-azobisisobutyronitrile
AMPSA	2-Acrylamido-2-methyl-1-propanesulfonic acid
ANOVA	analysis of variance
BOPP	biaxially oriented PP film
CAN	ceric ammonium nitrate
CAS-g-PAM	poly acrylamide grafted casein
CBD	chemical bath deposition
CHL	chloroform
CIE	Commission Internationale d'Eclairage
CIF 1	Central Instrumentation Facility
CPM	chlorpheniramine maleate
CSIR	Council of Scientific and Industrial Research
CVD	chemical vapor deposition
DFT	density functional theory
DMA/DMTA	dynamic mechanical (or thermal) analysis
DMF	N,N-Dimethyl formamide
DMSO	dimethyl sulphoxide
DOE	design of experiments
DSC	differential scanning calorimetry
DSMO	dimethylsulfoxide
DST	Department of Science and Technology
EG	ethylene glycol
EGDMA	ethylene glycol dimethacrylate
EL	electroluminescence
EMI	electromagnetic interference
FF	fill factor
FGF-2	fibroblast growth factor-2
FT-IR	fourier transform infrared spectrometer

FTO	fluorine-doped tin oxide
GLCM	gray-level co-occurrence matrix
GPC	gel permeation chromatography
HA	hydroxyapatite
HDPE	high density polyethylene
HTGPC	high temperature gel permeation chromatography
IA	itaconic acid
IPN	interpenetrating polymeric network
IR	infrared spectroscopy
KETEP	Korea Institute of Energy Technology Evaluation and Planning
LDPE	low density polyethylene
LINCS	linear constraint solver
LLDPE	linear low density polyethylene
MAA	methacrylic acid
MADM	multi attribute decision making
MD	molecular dynamics
MESER	Molecular Environmental Science and Engineering Research
MFI	melt flow index
MIPs	molecularly imprinted polymers
MMP-2	matrix metalloproteases-2
MRSA	methicillin-resistant Staphylococcus aureus
MSA	methanesulfonic acid
MSD	mean square displacement
MVR	melt volume rate
MWD	molecular weight distribution
NIPs	non-imprinted polymers
NMR	nuclear magnetic resonance
OFETs	organic field-effect transistors
OLEDs	organic light-emitting diodes
OPLS	optimized potentials for liquid simulations
OPVs	photovoltaic devices
PALS	positron annihilation lifetime spectroscopy
PAM	poly acrylamide
PAN	polyacrylonitrile

PANi	polyaniline
PBS	phosphate buffer solution
PC	poly carbonate
PCM	polarized continuum model
PEG	polyethylene glycol
PEI	polyethylenimine
PEN	polyethylene naphthalate
PET	poly ethylene terephthalate
PL	photoluminescence
PLED	polymer light emitting diodes
PME	particle mesh Ewald
PMMA/SAN	poly (methyl methacrylate)/poly (styrene acrylonitrile)
PP	polypropylene
PPC	pre-polymerization complex
PVC	poly vinyl chloride
PVP	poly(vinyl pyrrolidone)
S	sulfur
SAD	selected-area diffraction
SCF	self-consistent field theory
SCS	semiconductor characterization system
SDS	sodium dodecyl sulfate
SE	shielding effectiveness
SEM	scanning electron microscope
SILAR	successive ionic layer adsorption and reaction
TDS	total dissolved solid
TEM	transmission electron microscope
TSS	total suspended solid
USEPA	United States Environmental Protection Agency
VRE	vancomycin-resistant enterococci
WHO	World Health Organization
WVTR	water vapor transmission rate
XIS	xylene insoluble
XPS	X-ray photoelectron spectroscopy
XRD	X-ray diffraction
XS	xylene soluble

PREFACE

The field of advanced polymer materials has had the attention, imagination, and close scrutiny of scientists and engineers in recent years. This scrutiny results from the simple premise that, using building blocks with dimensions in the macro to nanoscale makes it possible to design and create new materials with unprecedented flexibility and improvements in their properties. The promise of nanocomposites lies in their multifunctionality, the possibility of realizing unique combinations of properties unachievable with traditional materials. The book is an attempt to cover the entire spectrum of advanced polymeric materials from macro to nano-length scales. We recognize that a book on a subject of such wide scope is a challenging endeavor and yet it is tried to introduce the recent research interests in the field of advanced polymeric materials.

The book entitled "*Advanced Polymeric Materials: From Macro- to Nano-Length* Scales" has 14 chapters. Chapter 1 deals with the polymer hydrogel dressing in wound management. Chapter 2 is about conducting solid polymer electrolyte membrane based on PVP and methanesulfonic acid. Chapter 3 outlines the novel multi-variable models for predicting the tensile and brittle strength of polymers. Chapter 4 discusses the dendritic organic semiconductors based on pyrene and triazine derivatives. In Chapter 5, rheological characteristics of LLDPE-fumed silica nanocomposites are discussed. Rational design of molecularly imprinted polymeric system is analyzed in Chapter 6. The EMI application of polymer/ferrite nanocomposites is outlined in Chapter 7. Chapter 8 is about the effect of speed ratio and cycle time on thickness of LLDPE products in rotational moulding process. Chapter 9 explains the molecular structure and property relationship of commercial biaxially oriented polypropylene. Chapter 10 deals with lead and cadmium ion removal by novel interpenetrating polymer-ceramic nanocomposites. Chapter 11 highlights the high performance polymeric flocculants based on modified milk protein—Microwave assisted synthesis. Chapter 12 is about polymer assisted synthesis of CdS nanostructure for photoelectrochemical solar cell applications and

synthesis and characterization chitosan-starch cross-linked beads is outlined in Chapter 13. Chapter 14 discusses the dendrimer polymer brushes
and its significance.

   We hope our readers will find the book of value to further their research
interests in this fascinating and fast developing area of advanced polymeric materials.

ABOUT THE EDITORS

Sabu Thomas, PhD

Sabu Thomas, PhD, is a Professor of School of Chemical Sciences and Honorary Director of the International and Inter University Centre for Nanoscience and Nanotechnology, Mahatma Gandhi University, Kottayam, Kerala, India. Since 1989 he has been associated with several universities in Europe, China, Malaysia, and South Africa. His research focuses include polymer blends, recyclability, reuse of waste plastics and rubbers, fiber-filled polymer blends, nanocomposites, elastomers, pervaporation phenomena, and sorption and diffusion. Professor Thomas is a member of the Royal Society of Chemistry of London, UK; a member of the New York Academy of Science, USA; and the recipient of awards from the Chemical Research Society of India and the Materials Research Society of India (2013). Prof. Thomas has supervised 65 PhD theses, and he has 17,500 citations to his credit. His *h*-index is 68.

Nandakumar Kalarikkal, PhD

Nandakumar Kalarikkal, PhD, is the Honorary Joint Director of the International and Inter University Centre for Nanoscience and Nanotechnology and Associate Professor in the School of Pure and Applied Physics of Mahatma Gandhi University, Kottayam, Kerala, India. Dr. Kalarikkal is also a visiting faculty member at Alemaya University and Mekkele University in Ethiopia, as well as a visiting fellow at the **Jawaharlal Nehru Centre for Advanced Scientific Research (JNCASR)** in Bangalore, India. His research interests include nanostructured materials, non-linear optics, laser plasma and phase transitions.

Maciej Jaroszewski, PhD

Maciej Jaroszewski, PhD, is an Assistant Professor and Head of the High Voltage Laboratory at Wroclaw University of Technology in Wroclaw, Poland. He received his MS and PhD degrees in high voltage engineering from the same university in 1993 and 1999 respectively. Dr. Jaroszewski was a contractor/prime contractor of several grants and a head of grant project on "Degradation processes and diagnosis methods for high voltage ZnO arresters for distribution systems" and is currently a contractor of a key project co-financed by the foundations of the European Regional Development Foundation within the framework of the Operational Programme Innovative Economy. His current research interests include high-voltage techniques, HV equipment diagnostics, HV test techniques, degradation of ZnO varistors, and dielectric spectroscopy.

Josmin P. Jose

Josmin P. Jose is a research scholar at the School of Chemical Sciences at Mahatma Gandhi University in Kottayam, Kerala, India. She is currently concentrating in organic/inorganic hybrid nanocomposites for her doctoral degree. She has completed her masters in chemistry from Loyola College, Chennai, and joined for research. Ms. Jose has published several research articles in high impact journals and a book chapter and has presented on "Hybrid Nanoparticle Based XLPE/SiO2/TiO2 and XLPE/SiO2 Nanocomposites: Nanoscale Hybrid Assembling, Mechanics and Thermal Properties" at an international conference held in Kuching, Malaysia, in September 2013. She is a co-editor of a book on polymer composites with Apple Academic Press, Inc. 2013. She is currently pursuing research experience with a very active polymer research group in INSA, Lyon, France.

CHAPTER 1

POLYMER HYDROGEL DRESSING IN WOUND MANAGEMENT

AJITH JAMES JOSE,[1] VINU VARGHESE,[2] and SAM JOHN[1]

[1]*Research and Postgraduate Department of Chemistry, St. Berchmans College, Changanacherry, Kerala 686 101, India*

[2]*Quality Control Department, Sance Laboratories Private Limited, Pala, Kottayam, Kerala 686573, India*

CONTENTS

1.1 Introduction 2
1.2 Polymer Hydrogel Dressings 2
1.3 Physical Characterization of Wound Dressings 3
1.4 Polymer Nanocomposite Hydrogels 8
1.5 Polymer hydrogels with Antimicrobial Activity 9
1.6 Chitosan in Wound Healing Applications 11
1.7 Alginate in Wound Dressing 12
1.8 Supramolecular Polymeric Hydrogels 13
1.9 Polymeric Drug Delivery Dressings 13
1.10 Future Perspectives 14
1.11 Conclusion 15
Keywords 15
References 16

1.1   INTRODUCTION

Wound care is one of the most lucrative and rapidly expanding medical device market for both manufacturers and providers. Wound management has recently become more complex because of new insights into wound healing and increasing need to manage complete wounds outside the hospital. The wound healing process is classically defined as a series of continuous, sometimes overlapping, events. These are hemostasis, inflammation, proliferation, epithelization, maturation, and remodeling of the scar tissue.

The last two decades have witnessed the introduction of many dresings, with new ones becoming available each year. These modern dressings are based on the concept of creating an optimum environment to allow epthelial cells to move unimpeded, for the treatment of wounds. Such optimum conditions include a moist environment around the wound, effective oxygen circulation to aid regenerating cells and tissues and a low bacterial load. Other factors, which have contributed to the wide range of wound dressings include the different type of wound (e.g., acute, chronic, exuding and dry wounds, etc.) and the fact that no single dressing is suitable for the management of all wounds. In addition, the wound healing process has several different phases that cannot be targeted by any particular dressing. Effective wound management depends on understanding a number of different factors such as the type of wound being treated, the healing process, patient conditions in terms of health (e.g., diabetes), environment and social setting, and the physical chemical properties of the available dressings. It is important therefore, that different dressings be evaluated and tested in terms of their physical properties and clinical performance for a given type of wound and the stage of wound healing, before being considered for routine use.

1.2   POLYMER HYDROGEL DRESSINGS

Polymer hydrogels are a still new, rapidly developing group of materials, gaining wide application in many fields, especially pharmacy, medicine and agriculture. Since the 1970s many reports have been published in the scientific literature about new chemical and physical structures, properties and innovative restricted applications of polymer hydrogels. Hydrogels are one of the simplest and most cost-effective method of removing devitalized tissue in wounds through autolytic debridement with wet dressings. On a molecular level, hydrogels are three-dimensional networks of hydrophilic polymers. Depending on the type of hydrogel, they contain varying percentages of water, but do not altogether dissolve in water.

Hydrogels can be applied either as an amorphous gel or as elastic, solid sheet or film. To prepare the sheets, the polymeric components are cross-linked so that they physically entrap water. The sheets can absorb and retain significant volumes of water upon contact with suppurating wounds. When applied to the wound as a gel, hydrogel dressings usually require a secondary covering such as gauze and need to be changed frequently. The sheets however, do not need a secondary dressing as a semi-permeable polymer film backing, with or without adhesive borders, controls the transmission of water vapor through the dressing. In addition the sheets can be cut to fit around the wound due to their flexible nature. The gels are used as primary dressings whereas the hydrogel films may be used as primary or secondary dressings. Hydrogel dressings contain significant amounts of water (70–90%) and as a result they cannot absorb much exudate, thus they are used for light to moderately exuding wounds.

Hydrogels possess most of the desirable characteristics of an 'ideal dressing.' They are suitable for cleansing of dry, sloughy or necrotic wounds by rehydrating dead tissues and enhancing autolytic debridement. These types of dressings are nonreactive with biological tissue, permeable to metabolites and are nonirritant. They also promote moist healing, are nonadherent and cool the surface of the wound, which may lead to a marked reduction in pain and therefore have high patient acceptability. Hydrogel gel dressing to treat a chronic leg ulcer for a patient who could not tolerate even reduced compression therapy due to pain, and the hydrogel helped reduce the pain considerably. Hydrogels also leave no residue, are malleable and improve reepithelization of wounds. In short, hydrogels are suitable for use at all four stages of wound healing with the exception of infected or heavily exuding wounds.

1.3 PHYSICAL CHARACTERIZATION OF WOUND DRESSINGS

The physical properties of all pharmaceutical formulations including wound dressings influence their ultimate performance and contribute to satisfying the desirable properties of dressings. The specific property to be characterized will depend on both the type of wound dressing, the nature of the surface to which the dressing will be applied and any secondary dressings that may be involved.

1.3.1   CHEMICAL/PHYSICAL ANALYSIS

The presence of different functional groups play an important role in the water holding capacity of the hydrogel, hence, it becomes necessary to analyze the presence of different functional groups in a newly synthesized hydrogel. Also, determination of the functional group can provide some information on the composition of the polymeric network. The various techniques, which are used for the purpose include infrared (IR) spectroscopy, UV-visible spectroscopy, nuclear magnetic resonance (NMR) and mass spectrometry.

The crystalline nature as well as structure of a substance has been determined using X-ray diffraction (XRD). The morphology and structure of the polymer hydrogels are analyzed using scanning electron microscope (SEM), atomic forced microscopy (AFM) and transmission electron microscope (TEM).

1.3.2   WATER VAPOR TRANSMISSION RATE

Water vapor transmission rate (WVTR) is defined as the quantity of the water vapor under specified temperature and humidity conditions, which passes through unit area of film material in fixed time. WVTR is measured in grams per square meter (g/m^2) over a 24 hours period according to the US standard ASTM E96–95. It is inversely proportional to the moisture retentive nature of a wound dressing, that is, the wound dressing with lower WVTR will be able to retain wound surface moisture. Typically, a wound dressing material showing WVTR less than 35 g/m^2/hr. is defined as moisture retentive and helps in a rapid healing.

1.3.3   MECHANICAL PROPERTIES

It is important to characterize the hydrogels for their mechanical properties. This is because the hydrogels could be used in various biomedical applications, viz. ligament and tendon repair, wound dressing material, matrix for drug delivery and tissue engineering, and as cartilage replacement, which requires hydrogels with different properties. FDA also provides strict guidelines for the same depending upon the type of applications. As for example, a drug delivery device should maintain its integrity during the lifetime of the application, unless it has been designed to degrade. A common method of increasing the mechanical strength is by increasing the crosslinking density, resulting in the formation of stronger gels, but with the increase in crosslinking density there is also a decrease in the % elongation of the hydrogels, that is, the hydrogels become brittle in nature. Hence, depending on the desired properties of the final products an optimum degree of crosslinking should be used. Copolymerization with a co-monomer, which may increase the H-bonding within the hydrogel, has also been utilized by many researchers to achieve desired mechanical properties.

1.3.4 HARDNESS

Hardness was determined using the hydrogel specimens, which were cut according to ASTM D-2240–95 with a thickness of 5–6 mm and tested with type A shore durometer. The measurements were carried out at 25°C and the data recorded 15 s after the pressing probe touched the specimen.

1.3.5 GEL FRACTION

The weight ratio of the dried hydrogels in rinsed and unrinsed conditions can be assumed as an index of degree of crosslinking or gel fraction. Therefore the gel fraction of samples can be calculated as follows

$$\text{Gel fraction}(\%) = \frac{W_f - W_c}{W_i - W_c} \times 100$$

where W_f and W_i are the weights of the dried hydrogel after and before rinsing and extraction, respectively and W_c is the weight of organoclay incorporated into the sample.

To perform gel fraction measurement, preweighed slice of each sample was dried under vacuum at room temperature until observing no change in its mass. Nearly identical weight of another slice of the same sample was immersed into excess of double distilled water for 4 days to rinse away unreacted species. Subsequently, the immersed sample was removed from double distilled water and dried at room temperature under vacuum until the dried mass showed constant weight.

1.3.6   ELASTIC MODULUS

This is the most fundamental and structurally important mechanical property of films and is a measure of film stiffness and rigidity. The elastic modulus is calculated from the slope of the initial linear portion of the stress-strain curve. A high elastic modulus indicates a hard, rigid film that is difficult to break. The elastic modulus, like the tensile strength is very sensitive to the presence of plasticizers such as water and glycerol. It has been used together with thermal techniques to determine the efficiency of different plasticizers in ethyl cellulose films. The elastic modulus is the major property employed during dynamic mechanical (or thermal) analysis (DMA/DMTA) for measuring the glass transition temperature for characterizing amorphous polymers and other polymeric dressings

1.3.7   RHEOLOGICAL TESTS

Dressings such as amorphous hydrogels can be characterized by measuring rheological properties such as viscosity and viscoelastic strength though the literature is very scanty. The rheological properties of rehydrated polymer dressings (gels) following gamma-irradiation in the glassy state have been investigated with a view to the suitability of gamma-radiation as a method of sterilizing such medicated dressings designed to remain on the surface of suppurating wounds for long periods of time. The quality of a hydrogel dressing comprising polyvinyl alcohol and polyvinyl pyrrolidone was investigated by measuring viscosity, gel fraction, glass transition temperature and water content. The authors noted that such properties when optimized could help meet ideal requirements of wound dressings such as absorbing fluid effectively, painless on removal, high elasticity and good transparency.

1.3.8 BIOADHESIVE STRENGTH

Another important physical property of dressings meant for application to moist wound surfaces, is adhesive strength both in vivo (bioadhesivity, mucoadhesion) and in vitro (adhesivity). Adhesivity has been defined as the force required to detach a sample from the surface of excised porcine skin (using a Texture Analyzer, a common type of mechanical testing equipment). The test is adopted from characterization of bioadhesive polymeric formulations meant for application to other moist surfaces such as vagina, buccal and nasal cavities. Adhesivity canalso be determined by evaluating various tensile responses of different gels. Adhesivity is important in wound healing where dressings should be self adhesive with the wound, easily removed and painless (i.e., it must have reduced adhesiveness with time). The force of adhesion depends on factors such as hydrophobicity, which is reported to improve bioadhesion, level of hydration and rate of polymer erosion in contact with the hydrating surface. A novel drug loaded wound dressing with optimized adhesive drug releasing properties was developed by binding self-adhesive Eudragit E (cationic copolymer based on dimethylaminoethyl methacrylate and neutral methacrylic esters) film with antibacterial loaded poly(N-isopropyl-acrylamide)microgel beads to achieve adhesive, absorptive and easy to peel functions.

1.3.9 BIOCOMPATIBILITY TESTS

Generally hydrogels are biocompatible and non-irritant in nature. The biocompatibility of the hydrogels is generally associated with the hydrophilic nature of the same, which helps in washing off the toxic and un-reacted chemicals during synthesis. The presence of water in the system makes it soft and rubbery, which offers least frictional irritation and provides a soothing effect when in contact with the physiological system. *In vitro* tests are not so expensive and hence are being preferred by the researchers to carry out the initial biocompatibility test.

In vitro tests for biocompatibility generally looks into the cyto-toxicity aspect of the material in the presence of live host cells and can usually done in two ways. In the first method, the material whose biocompatibility has to be determined is placed in direct contact with the host environmental cells and is subsequently incubated for a specific period of time at 37°C. In the second method, the material is placed in a suitable physiological solution and is incubated for a specified period of time at 37°C to allow any leaching from the material. The leachates, so obtained, are used to carry out the biocompatibility tests in the presence of the cells. The researchers usually determine the cell viability and cell proliferation from the cytotoxicity test. In a typical experiment, the hydrogels are usually sterilized (by immersing in ethanol and drying under Laminar air flow) followed by seeding with the host cells. After the seeding of the cells, the system is incubated for 1 h to allow cell attachment on the hydrogels which is followed by the addition of culture media and subsequent incubation at 37°C for allowing cell growth and proliferation. The cell proliferation can either be visualized by microscopy or by carrying out MTT (tetrazolium salt, 3-[4, 5-dimethylthiazol-2-yl]-2, 5-diphenyl tetrazolium bromide) assay. MTT assay is a colorimetric method, which allows quantification of cell growth and proliferation. The assay is based on the principle of reduction of MTT into purple colored Formosan crystals in the presence of mitochondrial dehydrogenase. These purple colored Formosan crystals are dissolved in dimethyl sulphoxide (DMSO) and are analyzed by measuring absorbance at 570 nm in a colorimeter. The quantity of Formosan crystals is directly proportional to the number of the live cells.

1.4 POLYMER NANOCOMPOSITE HYDROGELS

Recent advances in the chemical, physical, and biological fields combined with rising needs in the biomedical and pharmaceutical sectors have led to new developments in nanocomposite hydrogels for wound dressing applications. Nanocomposite hydrogels based on polyvinyl alcohol and organically modified montmorillonite clay were prepared by the cyclic freezing-thawing method, as novel wound dressings. The potential of PVA nanocomposite hydrogels for wound dressing applications were investigated by evaluating some of their important properties such as swelling and mechanical properties, also their ability in transmission of water vapor and resistance to microbe penetration. It was concluded that the quantity of clay was the key factor to obtain nanocomopsite hydrogels with desirable properties. According to the obtained results, the polyvinyl alcohol/clay nanocomposite hydrogels showed excellent physical and mechanical properties which met the essential requirements of ideal wound dressings. Based on swelling measurements, they exhibited high capability in absorbing fluid, so recommended for exudative wounds. Because of their unique mechanical properties, that is, very high elasticity, they could be excellent candidates for wounds under high stresses. The proper values of the water vapor transmission rates of polyvinyl alcohol nanocomposite hydrogels indicated that they could keep moist environment on interface of the wound and dressing to accelerate the healing process. Thus, polymer nanocomposite hydrogels are mechanically strong hydrogels able to absorb fluid exudate from wound sites in addition to preventing exogenous bacterial infiltration.

1.5 POLYMER HYDROGELS WITH ANTIMICROBIAL ACTIVITY

Critical colonization and infection of wounds present a dual problem for clinicians. First, there is the possibility of delayed wound healing, particularly in the presence of a compromised immune system or where the wound is grossly contaminated or poorly perfused. Second, colonized and infected wounds are a potential source for cross-infection-a particular concern as the spread of antibiotic-resistant species continues. For patients, an infected wound can have additional consequences including increased pain and discomfort, a delay in return to normal activities, and the possibility of a life-threatening illness. For healthcare providers, there are increases in treatment costs and nursing time to consider.

Silver has proven antimicrobial activity that includes antibiotic-resistant bacteria, such as methicillin-resistant *Staphylococcus aureus* (MRSA) and vancomycin-resistant enterococci (VRE). Its role as an antimicrobial agent is particularly attractive, as it has a broad spectrum of antimicrobial activity with minimal toxicity toward mammalian cells at low concentrations and has a less likely tendency than antibiotics to induce resistance due to its activity at multiple bacterial target sites.

Polymeric gels containing silver nanoparticles have been prepared by incorporating silver particles in situ by oxidation of the hydroxymethyl groups and reduction of $AgNO_3$ into hydrogel of poly(acrylamide)/poly(N-(hydroxymethyl)acrylamide). Silver nanocomposites of poly(acrylamide-co-acrylic acid) hydrogels were also synthesized involving the anion exchange between Ag^+ and H^+ and the subsequent reduction to Ag nanoparticles by trisodium citrate. The hydrogel shows good antimicrobial efficiency against E. coli, which is dependent, as expected, on the size and the amount of the nanoparticles. Silver has also been employed in multi-functional polymer based wound healing gel acting as gelation catalyst and antimicrobial agent. Moreover, silver nanoparticles were formed into polymer gels of polyether imide and hydroxyethyl acrylate upon reduction of $AgNO_3$. In addition, the network was modified with poly ethylene glycol, well known as antifouling polymer, combining both effects.

Researchers have paid special attention to graphene, a single layer and two-dimensional lattice with high mechanical strength, big surface area, good thermal and electrical conductivity and excellent biocompatibility. This is because, firstly, graphene can be produced from graphene oxide in a large-scale and low-cost strategy, which facilitates low-cost and large area preparation of graphene hydrogels. Secondly, graphene with high mechanical strength can be used as an efficient filler to enhance the mechanical properties of hydrogels. Thirdly, graphene hydrogels possess porous structure, large water absorption capacity and excellent biocompatibility, which makes graphene hydrogels conductive to cellular adhesion and growth. Fourthly, graphene hydrogels can maintain moist environment around wound thereby facilitating wound healing. This research on the wound dressing based on graphene hydrogel will extend new application range of graphene in biomedicine. Researchers at A STAR Singapore Institute of Manufacturing Technology and co-workers have now compared the antibacterial activity of graphite, graphite oxide, graphene oxide and reduced graphene oxide using the model bacterium Escherichia coli.

A series of Ag-graphene composite hydrogels have been prepared by crosslinking Ag/graphene composites with acrylic acid and N,N-methylene bisacrylamide in the presence of glucose as a green reducing agent, which is favorable for minimizing the toxicity effect on the tissues. It has been found that composite hydrogel with Ag to graphene mass ratio of 5:1 exhibits desired antibacterial abilities against both Gram-negative Escherichia coli and Gram-positive Staphylococcus aureus. Although these hydrogel contained more Ag than others, there was a negligible effect on the fibroblastic biocompatibility. In comparison with some reported hydrogels with high swelling ratio, higher swelling ratio can be achieved for the prepared hydrogel owing to the existence of partially reductive graphene and hydrophilic polyacrylic acid. Meanwhile, graphene embedded in hydrogel can enhance the tensile strength and elongation at break of hydrogel, and fit the mechanical necessary of dressing. When this hydrogel was used to cure full-thickness skin wound, a higher wound healing ratio in less time can be observed compared to other hydrogels and some reported dressings, and this result was further proved by histological analysis, demonstrating its potential application in wound treatment.

Dendritic structures based on polyglycerol and o-carboxymethylated chitosan have been synthesized, acquiring antimicrobial character after reaction with boric acid. Antibacterial studies of this polymer against *S. aureus* and *P. aeruginosa* showed, by measuring the inhibition zone diameter, a significant activity, almost double, in comparison to that of boric acid. In addition, its cytotoxicity was evaluated against Chinese hamster ovary cells by elution test. These dendritic polymers proved lower cytotoxicity than the controls (non-cytotoxic ultra high molecular weight polyethylene and cytotoxic phenol), suggesting that the material has improved biocompatibility. Moreover, the in vivo behavior was tested in female Wistar rats analyzing the healing process of wounds treated with dendritic polymer membranes. All the polymers are able to cure the injuries and also to regenerate the collagen fibers.

1.6 CHITOSAN IN WOUND HEALING APPLICATIONS

Chitosan-based materials, produced in varying formulations, have been used in a number of wound healing applications. Chitosan induces wound healing on its own and produces less scarring. It seems to enhance vascularization and the supply of chito-oligomers at the lesion site, which have been implicated in better collagen fibril incorporation into the extracellular matrix. While different material dressings have been used to enhance endothelial cell proliferation, the delivery of growth factors involved in the wound-healing process can improve that process. In addition to the reparative nature of the chitosan hydrogels they can also deliver a therapeutic payload to the local wound, for example, fibroblast growth factor-2 (FGF-2) which stimulates angiogenesis by activating capillary endothelial cells and fibroblasts. To sustain FGF-2 residence at the wound site, FGF-2 was incorporated into a high molecular weight chitosan hydrogel, formed by UV-initiated crosslinking.

A chitosan hydrogel scaffold impregnated with β-FGF-loaded microspheres were developed. That accelerates wound closure in the treatment of chronic ulcers. Films of chitosan, in combination were found to promote accelerated healing of incisional wounds in a rat model. The wounds closed within 14 days and mature epidermal architecture observed histologically with keratinized surface of normal thickness and a subsided inflammation in the dermis.

1.7   ALGINATE IN WOUND DRESSING

Alginate dressings in the dry form absorb wound fluid to re-gel, and the gels then can supply water to a dry wound, maintaining a physiologically moist microenvironment and minimizing bacterial infection at the wound site. These functions can also promote granulation tissue formation, rapid epithelialization, and healing. Various alginate dressings including Algicell™ (Derma Sciences) AlgiSite M™ (Smith & Nephew), Comfeel Plus™ (Coloplast), Kaltostat™ (ConvaTec), Sorbsan™ (UDL Laboratories), and Tegagen™ (3M Healthcare) are commercially available.

A variety of more functional and bioactive alginate based wound dressings have also been studied to date. The sustained release of dibutyryl cyclic adenosine monophosphate, a regulator of human keratinocyte proliferation, from partially oxidized alginate gels accelerated wound healing, leading to complete re-epithelialization of full thickness wounds within 10 days in a rat model. Alginate gels releasing stromal cell-derived factor-1 were also effective in accelerating wound closure rates and reducing scar formation in pigs with acute surgical wounds. Incorporation of silver into alginate dressings increased antimicrobial activity and improved the binding affinity for elastase, matrix metalloproteases-2 (MMP-2), and proinflammatory cytokines (e.g., TNF-α, IL-8). The addition of silver into alginate dressings also enhanced the antioxidant capacity. Alginate fibers cross-linked with zinc ions have also been proposed for wound dressings, as zinc ions may generate immunomodulatory and anti-microbial effects, as well as enhanced keratinocyte migration and increased levels of endogenous growth factors. Blends of alginate, chitin/chitosan, and fucoidan gels have been reported to provide a moist healing environment in rats, with an ease of application and removal.

1.8 SUPRAMOLECULAR POLYMERIC HYDROGELS

Supramolecular polymeric hydrogels, is a remarkable new class of soft and responsive materials, different from conventional covalent crosslinking methods. The combination of polymer chains and selective and strong supramolecular crosslinks presents an attractive platform from which one can readily modify structural parameters such as the polymeric backbone, the strength and dynamics of crosslinking interactions, directionality, multiple-responsiveness and tunable degradability in a modular fashion.

Supramolecular crosslinks allow for dynamic behavior: structural error correction, shear thinning, self-healing properties, elasticity and moldability. Many of the existing studies presented here were particularly focused on synthesis of supramolecular polymeric hydrogels. However, only very few studies have been emerging to describe completely and quantitatively the relationship between the fundamental parameters guiding self-assembly of the supramolecular crosslinking motifs and the macroscopic behavior of the resulting materials. Therefore, a need still exists to develop systematic studies on this complex relationship and the structure and dynamics of these supramolecular polymeric hydrogels in wound dressing. Nevertheless, the use of supramolecular chemistry in the assembly of networked structures has been demonstrated to allow the spatiotemporal control of the macroscopic viscoelastic properties of these hydrogels, which could have implications in biomedical applications. The current developments in this field have provided us with a toolbox from which we can utilize to build customizable structures and these advances bode well for the future development of improved supramolecular polymeric hydrogels.

1.9 POLYMERIC DRUG DELIVERY DRESSINGS

Most modern dressings are made from polymers, which can serve as vehicles for the release and delivery of drugs to wound sites. The release of drugs from modern polymeric dressings to wounds has been sparsely reported in the literature with few clinical studies carried out to date. The polymeric dressings employed for controlled drug delivery to wounds include hydrogels such as poly(lactide-co-glycolide), poly(vinyl pyrrolidone), poly(vinyl alcohol), and poly(hydroxyalkylmethacrylates) polyurethane-foam, hydrocolloid and alginate dressings. Other polymeric dressings reported for drug delivery to wounds comprise novel formulations prepared from polymeric biomaterials such as hyaluronic acid, collagen and chitosan. Synthetic polymers employed as swellable dressings for controlled drug delivery include silicone gel sheets, lactic acid. Some of these novel polymeric dressings for drug delivery exist as patents. Composite dressings comprising both synthetic and naturally occurring polymers have also been reported for controlled drug delivery to wound sites. Sustained release tissue engineered polymeric scaffolds for controlled delivery of growth factors and genetic material to wound sites have also been reported. The modern dressings for drug delivery to wounds may be applied in the form of gels, films and foams whilst the novel polymeric dressings produced in the form of films and porous sponges such as freeze-dried wafers or discs or as tissue engineered polymeric scaffolds.

1.10 FUTURE PERSPECTIVES

Several challenges remain that need to be taken into consideration in developing novel wound healing drug delivery formulations. For example, large variations in the rate of production of wound exudate, suggests the difficulty in finding a single ideal dressing capable of application to all wound types. It seems ideal to have composite dressings, which combine the different characteristics of current technologies. This will aid in targeting the many aspects of the complex wound healing process, to ensure effective, complete wound healing and shorter healing times for chronic wounds (and other difficult to heal wounds). It may also be expedient to employ individualized therapeutic approaches for treating specific wound types and individuals using emerging tissue engineering technologies. Such advanced approaches can help treat chronic wounds in a clinically efficient manner. However, large, randomized and controlled clinical trials to examine safety and efficacy will need to be carried out for many of these advanced dressings to speed up their use in routine clinical practice. There may be many unexplored polymeric dressings with idealized properties required for the effective and sustained delivery of therapeutic agents to chronic wounds.

1.11 CONCLUSION

Various hydrogels with improved structural properties and novel functions have already been prepared and explored for wound healing. The ability to engineer new classes of hydrogels with precisely controlled chemical and physical characteristics, unlike the limited repertoire available from natural sources, designed for a specific application could revolutionize the use of these materials. The knowledge of the cellular and molecular processes underlying wound healing has reached a level, which let researchers apply for new therapeutic approaches that act directly on cellular and subcellular events during the healing process. Similarly, nanotechnology today offers the means to overcome the dimensional barrier of currently used therapies for wounds and ulcers, to reach a dysfunctional molecular target and exert the therapeutic action straight at the origin of the chronic condition.

KEYWORDS

- atomic forced microscopy
- hemostasis
- nuclear magnetic resonance
- polymer hydrogels
- scanning electron microscope
- transmission electron microscope
- water vapor transmission rate
- wound management

REFERENCES

1. Boateng, J. S., Matthews, K. H., Stevens, N. E., Eccleston, G. M. Wound Healing Dressings and Drug Delivery Systems: A Review. *J. Pharm. Sci.* 97, 2892–2923 (2008).
2. Fan, Z. J., Wang, J. Q., Gong P. W., Ma, L. M., Yang S. R. A Novel Wound Dressing Based on Ag/Graphene Polymer Hydrogel: Effectively Kill Bacteria and Accelerate Wound Healing. *Adv. Funct. Mater.* 1–11 (2014).
3. Jian, W., Mather, P. T. Nanostructured Hydrogel Webs Containing Silver. *Biomacromolecules,* 10, 2686–2693 (2009).
4. Lee, K. Y., Mooney, D. J. Alginate: Properties and biomedical applications. *Prog. Poly. Sci.* 37, 106–126 (2012).

CHAPTER 2

AC IMPEDANCE AND FTIR STUDIES ON PROTON CONDUCTING POLYMER ELECTROLYTE MEMBRANE BASED ON PVP AND METHANESULFONIC ACID

C. AMBIKA,[1] G. HIRANKUMAR,[1] S. KARTHICKPRABHU,[1,2] and R. S. DARIES BELLA[1]

[1]C-SAR, School of Basic Engineering and Sciences, PSN College of Engineering and Technology, Melathediyoor, Tirunelveli, Tamilnadu, India; E-mail: g.hirankumar@psnresearch.ac.in

[2]Department of Physics, Kalasalingam University, Krishnankoil, Virudhunagar, Tamilnadu, India

CONTENTS

Abstract 18
2.1   Introduction 18
2.2   Experimental Details 20
2.3   Result and Discussion 21
2.4   Conclusion 28
Keywords 28
References 28

ABSTRACT

Solid polymer electrolytes are prepared by complexing methanesulfonic acid (MSA) with poly (vinyl pyrrolidone) (PVP) matrices. Various compositions of polymer electrolytes are tailored by solution casting technique using N,N-Dimethyl formamide (DMF) as solvent. The complex formation between the host polymer and methanesulfonic acid are conformed by FT-IR spectroscopy. Dependence of ionic conductivity with frequency is carried out at different temperatures. From impedance analysis, higher conductivity is found to be $9.72\pm0.03\times10^{-6}$ S/cm at ambient temperature for 16.77 mol% concentration of methanesulfonic acid. The enhancement of conductivity is observed to the order of 10^5 times with the addition of 16.77 mol% of MSA to PVP host polymer. At 20.82 mol% of MSA, decrease of conductivity is noticed and its value again increases gradually for 24.81mol%, 30.79 mol%, 34.65 mol% of MSA concentrations. Sudden drop of conductivity is again observed at 40.5 mol% of MSA. Temperature dependent dielectric relaxation and modulus spectra of PVP-MSA membranes are also studied using a.c. impedance spectroscopy.

2.1 INTRODUCTION

The field of solid polymer electrolytes is experiencing vigorous activities in recent years due to the significant theoretical interest as well as practical importance for the development of electro chemical devices such as energy conversion units (batteries/fuel cells), humidity and gas sensors, electrochromic display devices, photo electrochemical solar cells, etc. (Maccallum and Vincent, 1987; Owen et al., 1989; Armand, 1986; Ratner and Shriver, 1988). Current technologies are based on sulfonated membranes with Nafion as a foremost example. Such membranes have to be highly hydrated to be effective proton conductors. Protons on the SO_3H (sulfonic acid) groups "hop" from one acid site to another. When the temperature rises above 100°C, their conductivity drops dramatically (Gomez-Romero, 2005).

In order to achieve higher working temperature and higher conductivities, fabrication of novel proton conducting membranes has been the subject of continuous research. In recent years some new proton conducting polymer membranes have been successfully proposed. Okada et al. (2005) have prepared polymer membrane with PVA-PAMPS-PVP and attained the conductivity of about 0.088 S/cm. From the review of the literature, few work was reported by using PVP complexed with various inorganic acids or salts such as $AgNO_3$ (Jaipal Reddy, 1995), $KClO_4$ (Ravi, 2013), $NaClO_3$ (Naresh Kumar, 2001), $KBrO_4$ (Ravi, 2011), $LiClO_4$ (Rodríguez, 2013), IL (Saroj et al., 2013), $AgCF_3SO_3$ (Kim et al., 2004), NH_4SCN (Ramya et al., 2008), NH_4NO_3 (Vijaya et al., 2012) to make the polymer become ionically conductive. These inorganic acids or salts cause some serious drawback like corrosion on the electrodes.

Among many host polymers, PVP was chosen as a host polymer in the present work. This is due to the presence of carbonyl group (C=O) in the side chains of PVP and it forms a variety of complexes with various inorganic salts. It can also permit faster ionic mobility compared to other semi-crystalline polymers. PVP has high Tg (86°C) due to the presence of rigid pyrrolidone group. In the present work, the host polymer (PVP) is complexed with methanesulfonic acid (MSA) because it contains sulfonic acid groups. When methanesulfonic acid is dissociated it provides $CH_3SO_3^-$ (anion) and H^+ (cation), which is responsible for conduction.

Methanesulfonic acid is a clear, colorless, strong organic acid. The structure of MSA lends itself to many catalytic reactions because of its high acid strength and low molecular weight. MSA is used as an ideal catalyst for esterification because of good kinetics, ease of handling and biodegradability. Methanesulfonic acid is usually described as a "green acid" due to its environmental advantages (Gernon, et al., 1997; Baker et al., 1991). It has the melting point around −55°C. Aqueous MSA solution under normal atmospheric conditions is odor-free, due to the very low vapor pressure and evolves no dangerous volatiles, making it safe to handle. MSA is far less corrosive and toxic than the usual mineral acids employed in making polymer electrolytes and diverse industrial processes (Martyak and Ricou, 2004).

MSA has higher thermal stability. It can operate over a wide range of temperatures usually from 80°C to 150°C. MSA begins decompose at 180°C and the activation energy is 30Kcal/mol. Elisabeth Smela et al. (2005) reported the actuation performance of polyaniline in MSA. Alexandre

Oury et al. (1973) reported oxygen evolution on α-PbO$_2$ in methanesulfonic acid at different concentration. With the above background various compositions of PVP and MSA complexes are analyzed by using complex impedance spectroscopy and FTIR spectroscopy. Complex impedance spectroscopy is used to investigate electrical and dielectric properties of polymer electrolyte membrane.

2.2   EXPERIMENTAL DETAILS

Polyvinyl pyrrolidone (PVP) of molecular weight 40000(K_{30}) (spectrochem) and methanesulfonic acid (s d Fine-Chem) (LR) were used to prepare solid polymer electrolyte membranes. These membranes were prepared by means of solution casting technique. Appropriate weights of PVP and MSA were dissolved in N,N-Dimethyl formamide (merck) at different molar ratios are listed in Table 2.1. The solution was stirred at 70°C for 1 hour to obtain a homogeneous mixture. This solution is then cast on clean polypropylene dishes. Those dishes are kept in oven at 70°C to evaporate the solvent. After the complete evaporation of the solvent, freestanding films are produced and thickness was measured with the aid of a electronic digital micrometer screw gauge. FTIR studies were carried out with the aid of JASCO FT/IR-4100 Fourier transform infrared spectrometer in the wave region between 4000 and 550cm^{-1}. Ionic conductivity of the samples was determined by IM6 Zahner elektrik workstation.

TABLE 2.1   Sample Code and Their Composition of Polymer Electrolyte Membrane

Sample Code	Composition of Polymer electrolyte membrane
S1	95.75mol% PVP: 04.25mol% MSA
S2	91.56mol% PVP: 08.44mol% MSA
S3	83.23mol% PVP: 16.77mol% MSA
S4	79.18mol% PVP: 20.82mol% MSA
S5	75.19mol% PVP: 24.81mol% MSA
S6	69.21mol% PVP: 30.79mol% MSA
S7	65.35mol% PVP: 34.65mol% MSA
S8	59.50mol% PVP: 40.50mol% MSA

2.3   RESULT AND DISCUSSION

2.3.1   FTIR STUDIES

The complexation, nature and concentration of the various ionic species are important to understand the overall mechanism of conductivity in the polymer electrolyte. These interactions can induce changes in the vibrational modes of the molecules in the polymer electrolyte. FTIR spectroscopy is an important tool for the investigation of such polymer electrolyte membranes. Figure 2.1 represents the FTIR spectra of $PVP-CH_3SO_3H$ complexes of polymer electrolytes for only specified regions. For pure MSA, symmetric stretching of C-S bond will be observed at 779 cm^{-1}. This peak is found to be shifted in the polymer – acid complexes and is shown in Fig. 2.1 (a).

Peaks at 1036 cm^{-1} and 1126 cm^{-1} corresponds to $v_s(SO_3^-)$ and $v_s(S=O)_2$ vibrations respectively for pure MSA. In the polymer electrolyte membranes, above mentioned peaks are shifted to the region 1034–1063 cm^{-1} and 1157–1166 cm^{-1} for stretching vibrations of SO_3^- ions and $(S=O)_2$, respectively, which is shown in Fig. 2.1 (b).

This may be due to the interaction of SO_3^- ions with PVP. The peaks at 1316 cm^{-1}, 844 cm^{-1} of PVP are attributed to CH_2 wagging, CH_2 bending, respectively, and are also found to be shifted towards higher wave number side. The carbonyl group of PVP, which appears at 1649 cm^{-1}

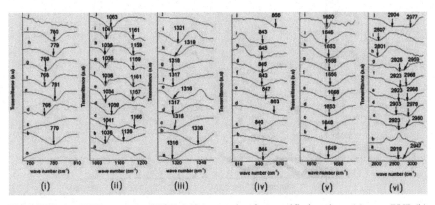

FIGURE 2.1   FTIR spectra of PVP-MSA complex for specified regions (a) pure PVP (b) pure MSA (c) S1 (d) S2 (e) S3 (f) S4 (g) S5 (h) S6 (i) 57 (j) S8 samples.

is also shifted to higher wave number side. This suggests that there is a strong interaction between C=O of pure PVP with cation of MSA. Figure 2.1(d) shows the peaks corresponding to symmetric CH_2 stretching at 2919 cm^{-1} and asymmetric CH_2 stretching at 2947 cm^{-1}. These CH_2 stretching peaks are also shifted towards high wave number side after doping of MSA. The above results confirm that there is a strong interaction between host polymer and MSA.

2.3.2 IMPEDANCE ANALYSIS

The typical impedance plot (Z' Vs. Z") for all the prepared polymer electrolyte membranes in the frequency range of 100 mHz to 300 KHz is shown in Fig. 2.2. The complex impedance diagram of some of the samples showed two regions: a semi-circle in the high frequency region which corresponds to bulk property of the membrane and the linear region in the low frequency range which is attributed to the effect of double layer capacitance at the blocking electrodes. The disappearance of semicircular portion in the high frequency region for few samples indicates that the conductivity is mainly due to ions.

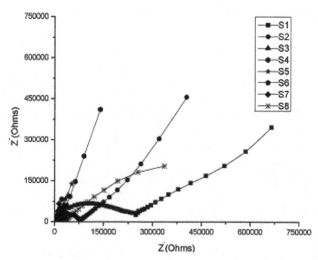

FIGURE 2.2   Cole-Cole plot for all samples at 303K.

Figure 2.3 shows the impedance plot for S3 sample, which has the higher conductivity among the prepared polymer electrolyte membranes. The bulk electrical resistance value (R_B) is calculated from the intercept on the real part of the impedance (Z') axis and the conductivity value can be calculated from the relation $\sigma = L/R_B A$ where L and A are the thickness and area of the sample. Conductivity values calculated by using the above relation for the prepared polymer electrolyte membranes are listed in Table 2.2.

2.3.3 VARIATION OF CONDUCTIVITY WITH METHANESULFONIC ACID CONCENTRATION

The ionic conductivity of the polymer electrolyte membrane as a function of MSA concentration is presented in the Fig. 2.4. It is observed that the conductivity value increases with the addition of MSA and it reaches a maximum value of 9.72×10^{-6} S/cm corresponding to 16.77mol% of MSA.

The enhancement of conductivity with increase in acid concentration upto 16.77 mol% may be due to the increase in concentration of mobile ions. At 20.82 mol% of MSA, the conductivity drops. This may be due to the formation of ion pairs ($CH_3SO_3^- H^+$). Till more addition of MSA upto 34.65 mol% results in raise of conductivity gradually. It happened due to formation of triplet ions

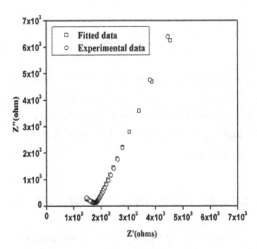

FIGURE 2.3   Cole-Cole plot for 83.23mol% PVP and 16.77mol% MSA at 303K.

TABLE 2.2   Conductivity Values at 303K and 323K for PVP Complexed with MSA Membranes

Sample Code	Conductivity at 303K (S/cm)	Conductivity at 323K (S/cm)
S1	1.09×10^{-7}	-
S2	2.48×10^{-7}	3.13×10^{-6}
S3	9.72×10^{-76}	2.43×10^{-4}
S4	1.43×10^{-6}	2.19×10^{-5}
S5	4.62×10^{-6}	8.96×10^{-5}
S6	5.55×10^{-6}	9.39×10^{-5}
S7	6.74×10^{-6}	-
S8	6.37×10^{-7}	3.51×10^{-5}

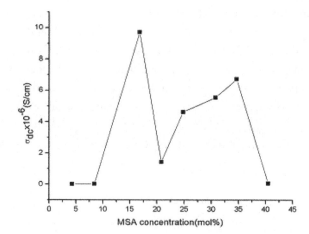

FIGURE 2.4   Effect of the concentration of MSA on the conductivity of PVP at 303K.

($H^+ CH_3SO_3^- H^+$). Steep decrease of conductivity is seen at 40.5 mol%, which may be due to the aggregation of MSA in the polymer matrix causes decrease of number of charge carriers.

2.3.4   TEMPERATURE DEPENDENT CONDUCTIVITY

As the temperature increases, polymer can expand easily and produce free volume. Thus ions, solvated molecules or polymer segments can move through the free volume and hence the conductivity increases [21]. The variation of conductivity with temperature can be elucidated with the help of Arrhenius relation, given below.

$$\sigma = \sigma_o \exp(-E_a/KT)$$

where σ_o is the pre-exponential factor, E_a is the activation energy in eV and T is the temperature in K. The activation energy values have been calculated from the slope of the Arrhenius plot. The values of activation energy for S2, S3, S4, S5 samples of PVP based MSA membranes are listed in Table 2.3. Among the four polymer electrolytes, S3 sample gives low activation energy, which suggests that only 0.96eV amount of energy is required to move the charge carriers.

In Fig. 2.5, Arrhenius plots for four samples (S2, S3, S4, S5) of PVP-CH$_3$SO$_3$H based polymer electrolytes in the temperature range of 303K-333K. Linear relations are observed because there is no phase transition in polymer matrix or the domain formed by the acid addition in the temperature range studied.

2.3.5 DIELECTRIC ANALYSIS

The dielectric behavior of the polymer electrolyte membrane were described by the real and imaginary parts of the complex permittivity $\varepsilon^*(\omega)$, which is defined by the relation, $\varepsilon^*(\omega) = \acute{\varepsilon}(\omega) - j\,\acute{\varepsilon}(\omega)$ where the real $\acute{\varepsilon}(\omega)$ and imaginary $\acute{\varepsilon}(\omega)$ components are the storage (dielectric constant) and loss of energy (dielectric loss) in each cycle of the applied electric field. Dielectric loss for S3 sample (shows higher conductivity) at different isotherms is represented in Fig. 2.6. The peak position of $\acute{\varepsilon}(\omega)$ shifts to

TABLE 2.3 Activation Energy Values For S2, S3, S4, S5 Samples

Sample code	Activation Energy E_a (eV)
S2	1.27
S3	0.96
S4	1.10
S5	0.97

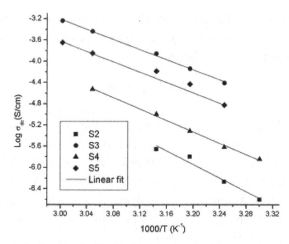

FIGURE 2.5   Arrhenius plot for S2, S3, S4, S5 polymer electrolyte membranes.

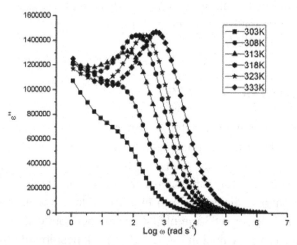

FIGURE 2.6   Variation of dielectric loss at different temperatures as a function of frequency for S3 sample.

higher frequency with increasing temperature. This high frequency shift with temperature suggests that a dielectric relaxation process occurs in the polymer electrolyte. Increase in peak value of $\acute{\varepsilon}(\omega)$ with the increase in temperature indicates an increase of charge carriers by thermal activation. At low frequencies, the alternating electric field is slow enough that the dipoles are able to keep pace with the field vibrations. As the frequency increases, $\acute{\varepsilon}(\omega)$ continues to increase and above the relaxation frequencies dielectric loss drop off as the electric filed is too fast to influence the dipole rotation and the orientation polarization disappears.

2.3.6 MODULUS ANALYSIS

In the modulus formalism, an electric modulus $M^*(\omega)$ is defined in terms of the complex dielectric permittivity $\varepsilon^*(\omega)$ as $M^*(\omega)=1/\varepsilon^*(\omega)= M'(\omega)+ j M''(\omega)$, where $M'(\omega)$ is the real and $M''(\omega)$ is the imaginary electric modulus. The variation of imaginary electric modulus with frequency for various temperature of S3 sample is presented in the Fig. 2.7. As the frequency increases, $M''(\omega)$ also increases. This increasing trend in the plot at higher frequencies may be attributed to the bulk effect. The height of $M''(\omega)$ peak decreases with rise in temperature, shows a plurality of relaxation mechanism. At very low frequencies, $M''(\omega)$ move towards zero, which suggests that there is a negligible contribution of electrode polarization. Long tail in the low frequency region gives the information about the large capacitance is associated with the electrodes.

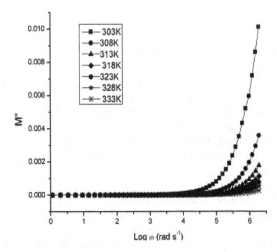

FIGURE 2.7   Variation of imaginary (M') part at different temperatures as a function of frequency for S3 sample

2.4 CONCLUSION

Polymer electrolytes based on PVP complexed with Methanesulfonic acid were prepared using solution casting technique. FTIR analysis confirms the complexation of MSA with host polymer. Higher conductivity of $9.72 \pm 0.03 \times 10^{-6}$ S/cm at ambient temperature is obtained for 16.77 mol% of methanesulfonic acid complexed with PVP. Conductivity shows Arrhenius behavior type of thermally activated process. Dielectric spectra indicate that the number of charge carriers may increase with the rise of temperature. Modulus analysis shows the non-Debye nature of the polymer electrolyte films.

KEYWORDS

- **methanesulfonic acid**
- **modulus analysis**
- **N,N-dimethyl formamide**
- **nafion**
- **poly (vinyl pyrrolidone)**

REFERENCES

1. Alexandre Oury, Angel Kirchev, Yann Bultd, Electrochim Acta 63 (2012) 28–36.
2. Armand, M. B., Ann. Rev. Mater. Sci. 16 (1986) 245.
3. Baker, S. C., Kelly, D. P., Murrell, J. C., Nature, 350 (1991) 627–628.
4. Elisabeth Smela, Benjamin Mattes, R., Synthetic metals, 151 (2005) 43–48.
5. Gernon, M. D., Wu, M., Buszta, T., Janney, P., Green Chem. 1 (1999) 127–140.
6. Jaipal Reddy, M., Sreepathi Rao, S., Laxminarsaiah, E., U. Subba Rao, V., Solid state ionics 80 (1995) 93–98.
7. Jinli Qiao, Takeo Hamaya, Tatsuhiro Okada, Polymer 46 (2005) 10809–10816.
8. Jong Hak Kim, Byoung Ryul Min, Chang Kon Kim, Jongok Won, Yong Soo Kang, J Polymer Science: Part B: Polymer Physics, 42 (2004) 232–237.
9. Jorge Rodríguez, Elena Navarrete, Enrique Dalchiele, A., Luis Sánchez, José Ramón Ramos-Barrado, Francisco Martín, J of Power Sources, 237 (2013) 270–276.
10. Maccallum, J. R., C. A. Vincent (Eds.), Polymer electrolyte reviews, Elsevier, Amsterdam, 1987.

11. Martyak, N. M., Ricou, P., Mater. Chem. Phys. 84 (2004) 87–98.

12. Miyamoto, T., Shibayama, K., J Appl. Phys. 44 (1973) 5372.

13. Naresh Kumar, K., Sreekanth, T., Jaipal Reddy, M., U. Subba Rao, V., J of power sources, 101 (2001) 130–133

14. Owen, J. R., Lasker, A. L., S. Chandra (Eds.), Superionic solids and solid electrolytes-Recent trends, Academic press, New York, 1989.

15. Pedro Gomez-Romero, Juan Antonio Asensio, Salvador Borros, Electrochim. Acta 50 (2005) 4715–4720.

16. Ramya, C. S., Selvasekarapandian, S., Hirankumar, G., Savitha, T., Angelo, P. C., J Non-Cryst. Solids, 354 (2008) 1494–1502.

17. Ratner, M. A., Shriver, D. F., Chem. Rev. 88 (1988) 109.

18. Ravi, M., Bhavani, S., Y.Pavani, V. V. Narasimha Rao, R., Ind J pure and appld physics 51 (2013) 362–366.

19. Ravi, M., Y.Pavani, Kirankumar, K., Bhavani, S., Sharma, A. K., V. V. Narasimha Rao, R., Mater. Chem. Phys. 130 (2011) 442–448.

20. Saroj, A. L., Singh, R. K., Chandra, S., Materials Science and Engg. B, 178 (2013) 231–238.

21. Vijaya, N., Selvasekarapandian, S., G.Hirankumar, Karthikeyan, S., Nithya, H., Ramya, C. S., Prabu, M., Ionics, 18 (2012) 91–99.

CHAPTER 3

DENDRITIC ORGANIC SEMICONDUCTORS BASED ON PYRENE AND TRIAZINE DERIVATIVES

RENJI R. REGHU and JUOZAS V. GRAZULEVICIUS

Department of Polymer Chemistry and Technology, Kaunas University of Technology, Radvilenu pl. 19, LT-50254 Kaunas, Lithuania, Tel: +37037 300193; Fax: +37037 300152; E-mail: juozas.grazulevicius@ktu.lt

CONTENTS

Abstract 31
3.1 Introduction 32
3.2 Dendrimers for Organic Electronics:
Structure-Property Relationship 33
3.3 Dendritic Pyrene Derivatives for Organic Electronics 35
3.4 Triazine-Core Derived Dendritic Compounds for
Organic Electronics 43
3.5 Conclusion 54
Acknowledgment 55
Keywords 55
References 56

ABSTRACT

In comparison to low-molar-mass compounds or polymers, dendritic organic materials constitute a superior class of organic semiconductors because of their distinctive physical properties relevant to application in organic electronics. Taking account on two important classes of functional dendritic cores, that is, pyrene (electron-rich) or triazine (electron-deficient) moieties, this review article concisely discusses the impact of peripheral chromophore-substitutions on their charge-transport, photophysical, electrochemical and thermal properties, and device performance.

3.1 INTRODUCTION

Electronic devices using organic materials as active layer(s), for example; electrophotographic photoreceptors, organic light-emitting diodes (OLEDs), organic photovoltaic devices (OPVs) or organic field-effect transistors (OFETs), have recently received enormous research interest from both academia and industry from the standpoint of potential technological applications as well as fundamental sciences. Organic materials-impelled devices are attractive because they can take advantage of organic materials such as lightweight, potentially low cost, capability to form large-area flexible thin-films. Photoreceptors in electrophotography using organic photoconducting materials have already established wide markets of copying and laser printers (Murayama, 1999; Kukuta, 1990). OLEDs have also found practical applications in lighting and displays. OPVs are viewed as one of the most promising candidates for cost-effective energy sources because of the possibility of a production on flexible and large-area substrates by solution processing that should radically reduce the manufacturing costs. OFET-driven displays have also been applied in low-cost electronics like logic circuits for radiofrequency identification tagging, smart cards or chemical and biological sensors.

Charge-transport is the essential operational process in the aforementioned organic electronic devices. Therefore, extensive development of high performance charge-transporting materials (organic semiconductors) is a key issue for employing them in competitively performed electronic devices. According to the charge-carriers that can be transported, charge-transporting materials can broadly be classified into three types. They are *p*-type organic semiconductors (hole-transporting), *n*-type organic semi-conductors (electron-transporting) and ambipolar organic semiconduc-tors (capable of transporting both holes and electrons). They are mostly based on π-electron systems and characterized by properties such as light absorption and/or emission in the ultraviolet-visible-near infrared wave-length region, charge-carrier generation and/or transport, nonlinear optical properties, etc.

Organic charge-transporting materials include small molecules (i.e., molecular materials), dendrimers and polymers. Charge-transporting materials having dendritic architecture are particularly interesting since their dimensional features, facile processability, monodisperse character-istics and the extended conjugation brings unique thermal, electrochemi-cal, photophysical and photoelectrical properties in addition to the tunable charge-transport which are decisive for device fabrication.

Alternatively, charge-transport is not only depending on the intrinsic electronic properties of the materials but also rely on the microscopic as well as macroscopic order of the materials in solid state. Hence, cutting-edge dendritic architecture of charge transporting materials can robustly impose extra order from nano to macro length scale in solid state by mostly exploiting their inherited supramolecular properties. Indeed, increased order can largely facilitate the charge carrier transport through the molecular segments and thus, considerably improve or modify the device performance.

3.2 DENDRIMERS FOR ORGANIC ELECTRONICS: STRUCTURE-PROPERTY RELATIONSHIP

Dendrimers are repeatedly branched and roughly spherical large mol-ecules. The term dendrimer is derived from the Greek words: *dendrons* –tree and *meros* –part. A dendrimer is typically symmetric around the core, and often adopts a spherical three-dimensional morphology (Fig. 3.1). It usually consists of three components: (1) the core which is located in the geometrical center or focus and usually determines the most important function of the dendrimer, for example, semiconducting characteristics; (2) the surrounding dendrons which contain branching points that define the dendrimer generation are particularly important to identify the effec-tive π-conjugation; (3) the surface groups which are covalently grafted

The Dendritic Stucture

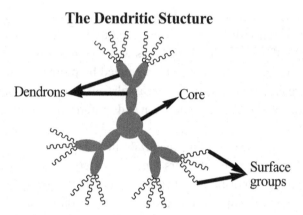

FIGURE 3.1   Schematic representation of dendritic structure.

onto the periphery of the dendrons with the intention of manipulating the solubility and processability of the dendrimer. The dendritic architectural feature allows a variety of possible combinations of core, dendrons and surface groups, which facilitate fine-tuning of each of the three functional parts independently without sacrificing others.

Dendritic organic semiconductors can be prepared by symmetrical multi-functionalization of polycyclic or heterocyclic aromatic core by various electro-active molecular moieties. Both divergent and convergent protocols are usually employed for the synthesis of dendrimers. However, convergent methodologies are more practiced for the preparation of dendritic organic semiconductors since this constructive building-block approach affords fine-tuning of the properties, which are crucial for high-performance devices (Turnbull and Stoddart, 2002).

Due to the careful structural design, dendrimers can combine the potential advantages of both small molecules and polymers (Astruc et al., 2010). Like small molecules, dendrimers possess a well-defined structure and monodispersity. This ensures batch-to-batch reproducibility unlike the polymers in terms of monodispersity and chemical purity. Both of them are essential factors for determining the device performance. At the same time, superior solubility and appropriate viscosity of dendrimers in common organic solvents make them suitable for solution processing by employing cost-effective methods like spin-coating or ink-jet printing to fabricate thin films for devices. Thus, expensive and high-temperature vacuum evaporation technique, which is applied commonly for small molecules can be evaded. Moreover, dendrimers containing photo-active components, in the core and/or in the branches, are particularly interesting since: (i) luminescence signals tender a handle to better understand the dendritic structures and superstructures; (ii) cooperation among the photoactive components can allow the dendrimer to execute useful functions such as light harvesting; (iii) changes in the photophysical properties can be exploited for sensing purposes with signal amplification; and (iv) photochemical reactions can modify the structure and other vital properties of dendrimers. These advantages of dendrimers over polymers as well as low-molar mass compounds, of course, inspire the synthetic strategies of chemists working on materials relevant to organic electronics.

Synthetic interest emphasized topics of dendrimers are already well addressed early on and continuously reviewed for nearly two decades. Moreover, the spectrum of organic semiconductors possessing dendritic shape is very broad. Hence, in this review article, we will be unambiguous in terms of structure-properties relationship of dendrimers that determines the functional characteristics of organic semiconductors. By calling attention to widely known building blocks/scaffolds for the synthesis of organic semiconductors like pyrene or triazine moieties as respective examples of electron -rich or –poor cores, correlation of dendritic structure to their decisive parameters significant to device applications will be discussed.

3.3   DENDRITIC PYRENE DERIVATIVES FOR ORGANIC ELECTRONICS

Pyrene is a flat aromatic system consisted of four fused benzene rings and termed as smallest peri-fused (one where the rings are fused through more than one face) polycyclic aromatic hydrocarbon. Since pyrene is peri-fused, aromatic core is well resonance-stabilized; which consecutively enhance effective π-electron delocalization. Electron-donor functionalized pyrene derivatives, with their extended delocalized π-electron systems, discotic shape, high photoluminescence efficiency, and good hole-injection/transport properties, have the potential to be a very interesting class of organic semiconductors for electronic applications.

Most of the dendritic materials based on pyrene core are tetrafunctional derivatives. Z. Zhao and et al. (2008) reported on pyrene core based stiff dendrimers containing carbazole/fluorene electron-donors as dendrons for application in OLEDs (Fig. 3.2). Pd/Cu-catalyzed Sonogashira coupling reaction (Sonogashira et al., 1975) has been employed as a key reaction to construct these acetylene-linked dendrimers (**Ia-Ie**). As demonstrated by

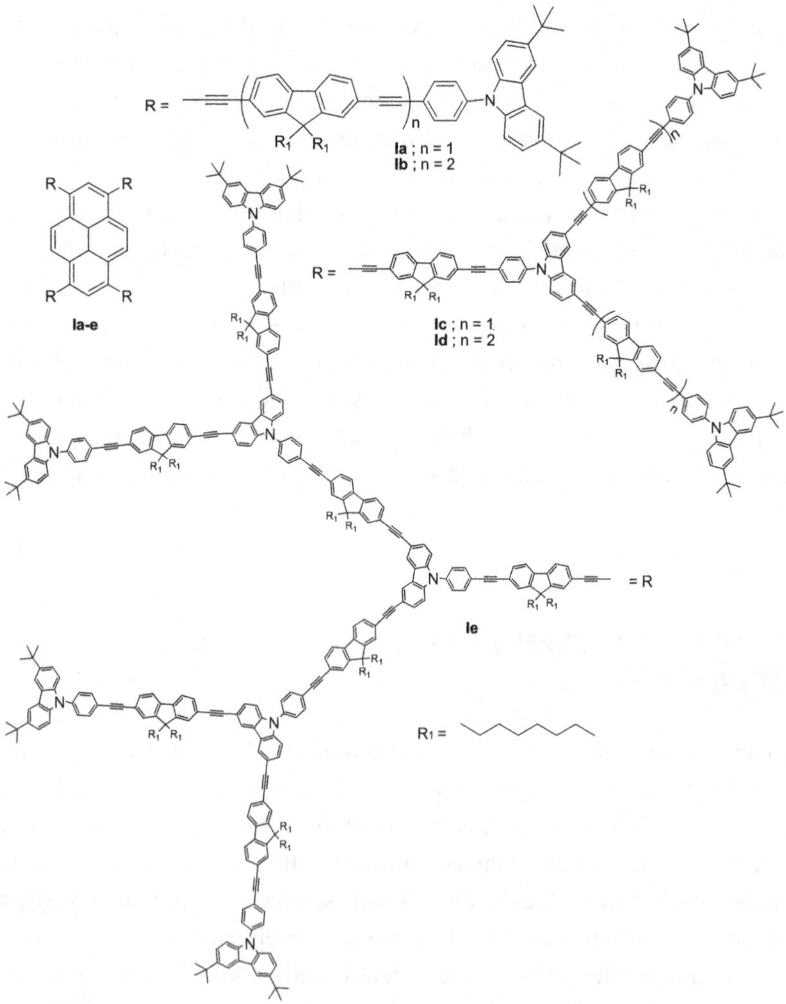

FIGURE 3.2 Ia-e.

theoretical method, the dendrimers **Ia** and **Ib** exhibited an 'X' shape with a calculated length of *ca.* 5 and 7 nm, respectively, and **Ic, Id** and **Ie** showed a dense "pancake-like" shape with a calculated diameter of *ca.* 9, 11, and 14 nm, respectively. The dendrons of **Ie** were found to be comparatively distorted in order to avoid the involved steric crowdedness that attributed to the high number of arms at its periphery. Therefore, the whole molecule was "thick" in the periphery and "thin" in the center, which hampered the pyrene ring from another molecule to get close to form excimers.

Dendrimers **Ia-e** exhibited high glass transition temperatures and enjoyed high thermal stability with decomposition temperatures in the range of 425–442°C. Lower generation dendrimers **Ia-Ib** were highly fluorescent in both solutions and solid states. The photolumines-cence (PL) quantum yields of the dendrimers were much higher than that of 1,3,6,8-tetrakis(phenylethynyl) pyrene (Venkataramana and Sankararaman, 2005). The quantum yield was found to be decreased with increase of dendrimer generation. Energy transfer efficiency might be decreased as the generation increases, since some of the arms were far away from the core (Devadoss, et al., 1996). The dendrimer generation effect on electroluminescence (EL) was tested in single-layer devices with a configuration of ITO(120nm)/PEDOT(25nm)/dendrimer/Cs_2CO_3(1nm)/ Al (100nm) (Huang, et al., 2006). **Ia** and **Ib** showed greenish yellow EL at 576 and 572 nm with Commission Internationale d'Eclairage (CIE) chromaticity coordinates of (0.48, 0.51) and (0.47, 0.52), respectively. **Id** showed yellow EL (CIE: 0.51, 0.49) at 560 nm. However, **Ie** exhibited a narrow yellowish green (CIE: 0.39, 0.59) EL at 530 nm with a weak emis-sion at *ca.* 560 nm, which was comparable to its photoluminescence in both solutions and solid state. **Ic**-based OLED exhibited yellow EL (CIE: 0.49, 0.50) with a maximum brightness of 5590 cd/m^2 at 16 V and a high current efficiency of 2.67 cd/A at 8.6 V. Gigantic two-photon absorption and strong two-photon excited fluorescence in these dendrimers have also been reported (Wan et al., 2005).

Sonogashira coupling reaction was also employed for the synthesis of cruciform-shaped pyrene derivatives **IIa-d** (Fig. 3.3) (Hu et al., 2010). Though **IIa-c** and **IId** contained same numbers of *p*-functionalized phenyl-ethynyl groups, their molecular structures were quite different as a result of phenylethynyl substituents at different positions of pyrene scaffold.

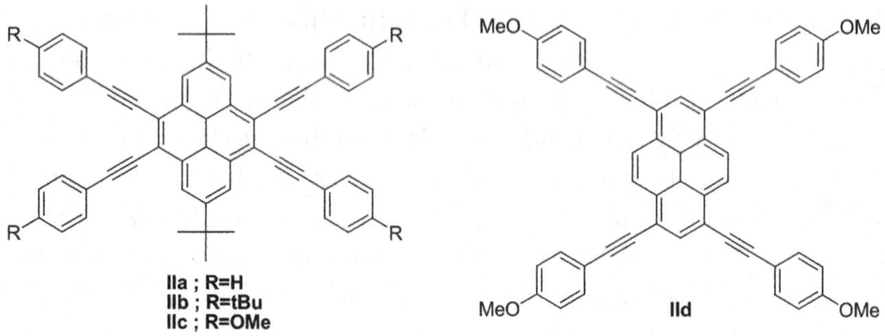

IIa ; R=H
IIb ; R=tBu
IIc ; R=OMe

IId

FIGURE 3.3 IIa-d.

Isomers **IIc** and **IId** were also different in the length and shape of their conjugation pathways. Inspection of absorption and emission spectra of **IIa-d** indicated that the extension of π-conjugation in pyrene chromophores lead to the shift of the wavelengths of absorption and fluorescence emission into the pure-blue visible region. Single-crystal X-ray analysis indicated that two bulky *t*-Bu groups on the pyrene moiety at the 2- and 7-positions played a significant role in inhibiting the π-stacking interactions between neighboring pyrene units. **IIa-d** emitted very bright, pure-blue fluorescence and showed good solubility in common organic solvents.

A series of compounds for OFETs, with pyrene moiety at the core and four substituted thiophene arms of different lengths (1–3 thiophene units) was reported (Fig. 3.4) (Anant et al., 2010). Pyrene functionalized at the 1,3,6,8-positions with unsubstituted thiophene units has also been reported (Ashizawa et al., 2008), but suffered from poor solubility. The presence of alkyl-chains at the termini of all oligothiophene substituents in **IIIa-c** remarkably improved the solubility in organic solvents that permitted ready characterization and processability. Absorption and PL spectra of compounds **IIIa-c** studied for both solutions and thin films showed red-shifted absorption and emission maxima as the oligothiophene length increased. X-ray solid-state study of **IIIa** demonstrated a layered structure with regions of alkyl chains and thiophene/pyrene units. Despite to this segregation, intermolecular contacts between the thiophene and pyrene components were found to be limited to edge-to-face interactions without the involvement of sulfur atoms in of thiophene rings. Hole transporting (p-channel) field-effect transistors based on drop cast films of compound **IIIc** showed a

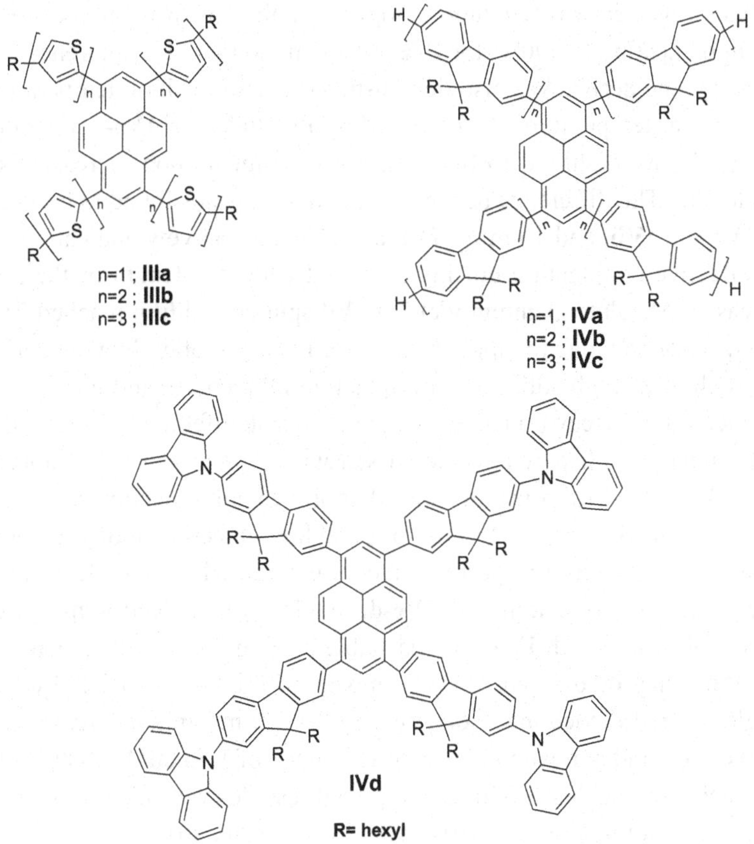

n=1 ; **IIIa**
n=2 ; **IIIb**
n=3 ; **IIIc**

n=1 ; **IVa**
n=2 ; **IVb**
n=3 ; **IVc**

IVd

R= hexyl

FIGURE 3.4 IIIa-c.

charge mobility of 1.9×10^{-5} cm^2/Vs under nitrogen environment. This hole-mobility value was lower than that of single crystals or films of the related tetra(2-thienyl)pyrene derivatives (Ashizawa et al., 2008).

Liu et al. (2008) reported pyrene centered oligofluorenes **IVa-c** with a good film forming ability and sky blue fluorescence. Carbazole end-capped pyrene derivative **IVd** possessing enhanced electrochemical stability and good device performance was also reported by the same research group. **IVa-c** showed high thermal stability with T_D ranging from 377 to 391°C. Thermal stability of **IVa-c** found to be increased as the oligofluorene chain length increased. T_D of **IVd** was lower than that of the oligofluorene-armed pyrene derivatives. Compared to the short arm species **IVa-b**, DSC analysis revealed a better morphological stability with no phase transition upon heating to 300°C for **IVc** possessing longer oligofluorene arms. These pyrene derivatives also demonstrated glass transition temperatures above room temperature. As observed by PL studies of **IVa-c**, the emission peaks were shifted to the longer wavelengths upon increasing the arm length. The difference between the emission peaks found to be 11 nm for **IVa** and **IVb**, and 1 nm for **IVb** and **IVc**, respectively, indicating that the effective conjugation quickly saturated when the length of the arms increased. Absolute quantum yield of **IVd** spin-coated film reached 71%, which is a good value for application in OLEDs. The oligofluorene derivatives exhibited high solid-state PL quantum efficiencies and near single-exponential PL decay transients with excited state lifetimes of ~1.4 ns.

Compared with pyrene-centered starbursts containing oligofluorenes arms, **IVd** with carbazole unit showed improved hole-injection ability and electrochemical stability. Because of good hole-injection ability, no additional hole-injection/transporting layer was required for the electroluminescent device fabrication with **IVc-d**. Single-layered electroluminescent device fabricated with **IVd** showed stable blue emission with a peak current efficiency of 0.84 cd/A and a maximum brightness of 2200 cd/m². Single-layered device made of **IVc** possessed a maximum brightness of 2700 cd/m^{-2} and a maximum current efficiency of 1.75 cd/A. Distributed-feedback lasers with low optical pumping thresholds were also constructed on the base of these pyrene derivatives (Sonar et al., 2010).

Sonar et al. (2010) reported soluble pyrene based organic semiconductors **Va-c** where electron-rich chromophores like phenylene, thienothiophene or benzothiadiazole-thiophene were employed as tetra substituents (Fig. 3.5). To achieve solution processability, alkyl groups were incorporated into the conjugated moieties. Solution PL spectra for **Vb** and **Vc** showed deep blue and sky blue emission, respectively, with the maxima at 433 and 490 nm, whereas **Va** exhibited orange emission with the maximum at 541 nm. The calculated HOMO values were found to be in the range of –5.15 to –5.33 eV, which matched quite well with commonly used hole injection/transport materials and anodes such as PEDOT:PSS (–5.1 eV) and ITO (–4.9 eV). The thermal decomposition temperatures were observed above 400°C for all the compounds except **Vb**, which showed thermal degradation onset at *ca.* 299°C. OLED based on **Vc** as

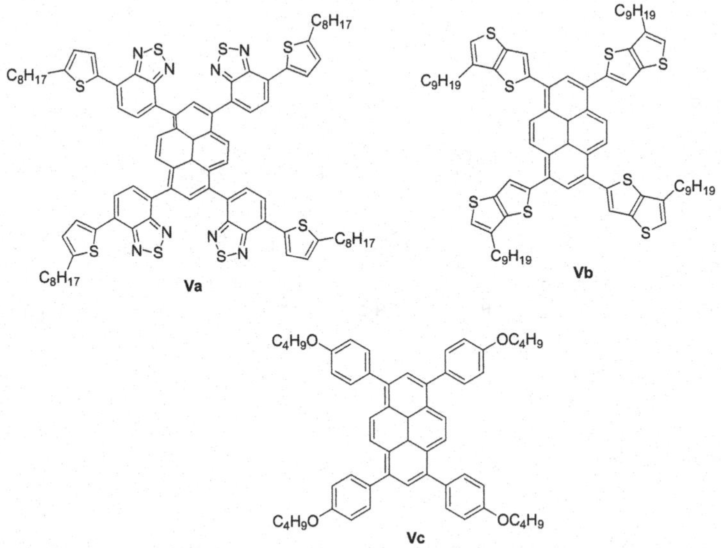

FIGURE 3.5 Va-c.

the emitter showed efficiency of 2.56 cd/A and a maximum brightness of 5015 cd m^{-2} at 11 V, and was characterized by deep blue emission (CIE: 0.15, 0.18) and low turn-on voltages (3.0 V).

Dendrimers **VIa-b** consisted of a polysulfurated pyrene core appended with poly(thiophenylene) dendrons possessing remarkable luminescent and electrochromic properties (Fig. 3.6) [43]. Photophysical and redox properties of these dendrimers were largely dependent on the length of their branches: (i) the dendron localized absorption band at *ca.* 260 nm increased strongly in intensity and moved slightly to the red on increasing dendrimer generation; (ii) the quantum yield, lifetime of the fluorescence and the values of half-wave potentials increased with the increase of dendrimer generation; (iii) the dendrimer branches partially protected the core from oxidation.

Using 1,3,6,8-tetraethynylpyrene as a core, a series of polyphenylene dendrimers was prepared by combining both divergent and convergent growth methods (Bernhardt et al., 2006). Fluorescence quantum yields of dilute solutions of these derivatives were found to be high (*ca.* 90%) and independent of the size of the polyphenylene shell.

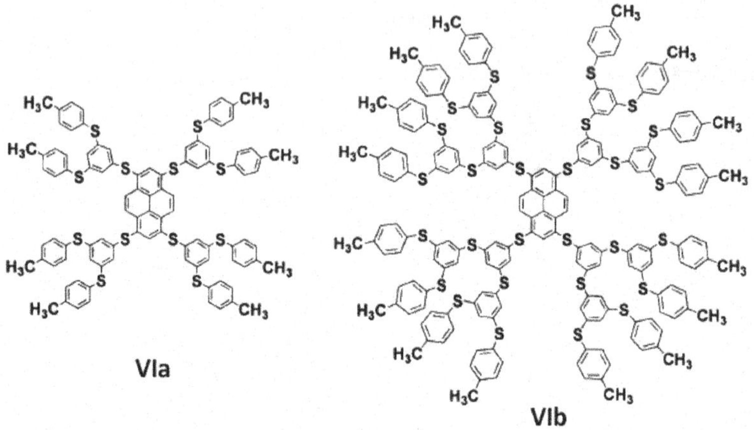

FIGURE 3.6   VIa-b.

Glass-forming pyrene-core based derivatives **VIIa-d** with the arms of differently linked carbazoles or phenothiazines were reported (Fig. 3.7), and their relevant properties for device applications were studied (Reghu et al., 2012). **VIIa-d** formed molecular glasses with high glass transition temperatures. Carbazole arms in the tetra functionalized pyrene derivatives were found to be twisted with respect to the pyrene core; however, the molecules tended to achieve planarization in the excited state, thus increasing effective conjugation and fluorescence quantum yields up to 0.84 in dilute solutions. They also displayed high fluorescence quantum yields (up to 0.60) when molecularly dispersed in polymer films at low concentration. Meanwhile, the neat films showed fluorescence decay time shortening (in the initial stage) and considerable drop in fluorescence quantum efficiency that indicated exciton-migration-induced quenching at non-radiative decay sites. Redox-active carbazolyl-substituted pyrene derivatives exhibited dicationic behavior and underwent electro-polymerization. Ionization potentials of thin layers of these materials were ranged from 5.2eV to 5.5 eV. Carbazolyl-substituted pyrene derivative **VIIc** transported holes with the drift-mobility of 5.8×10^{-5} cm^{2}V^{-1}s^{-1} at an electric field of 10^6 Vcm^{-1} as characterized by xerographic time of flight technique. However, when electron-donors like carbazole or fluorene were employed as central chromophore, pyrene derivatives emitted in the blue region and exhibited enhanced photophysical properties.

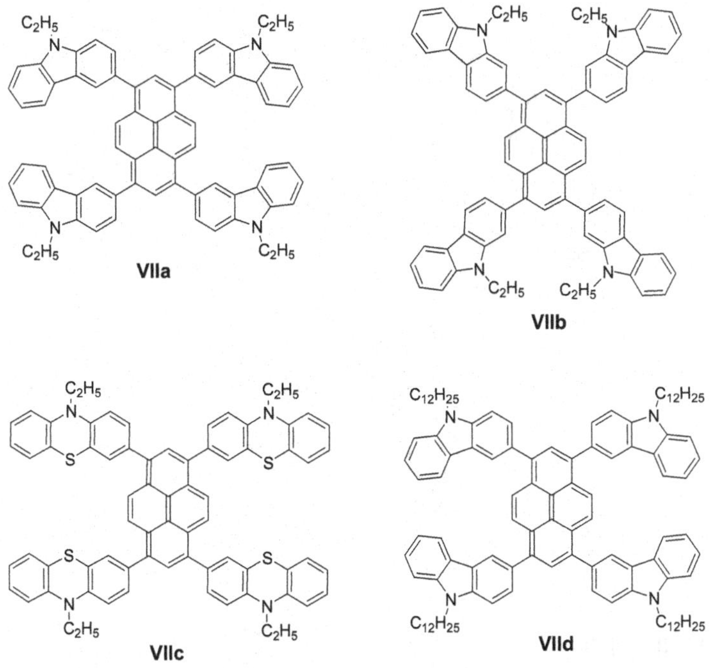

FIGURE 3.7 VIIa-d.

3.4 TRIAZINE-CORE DERIVED DENDRITIC COMPOUNDS FOR ORGANIC ELECTRONICS

Triazine is an electron-deficient heterocyclic ring, analogous to the six-membered benzene ring but with three carbon atoms replaced by nitrogen atoms. 1,3,5-Triazine (s-triazine) derivatives have proven their great potential in the area of material chemistry, both for their π-interaction abilities, and for their aptitude to be involved in intricate H-bond networks. Because of the electron-poor characteristics, most of the triazine derivatives are reportedly known as n-type organic semiconductors. However, combination of electron-accepting triazine chromophores and electron-donating species can strategically produce materials exhibiting ambipolar charge-transport.

Electron transporting host materials **VIIIa-c** were reported for green phosphorescent OLEDs based on s-triazine and various aryl moieties (Fig. 3.8). Thermal, photophysical and charge-transporting properties as

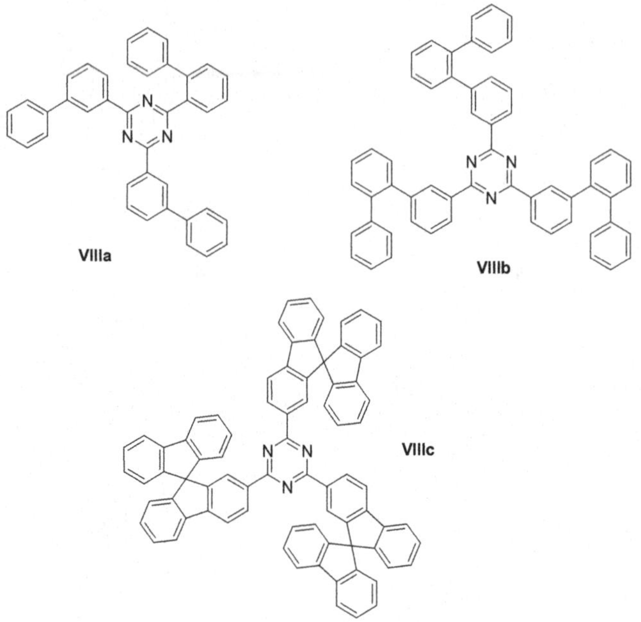

FIGURE 3.8　VIIIa-c.

well as morphology of **VIIIa-c** found to be influenced by the nature of the aryl substituents attached to the triazine core. The *meta–meta* linkage between the triazine core and the peripheral aryl moieties in **VIIIa-b** restrict the effective extension of their π-conjugation that lead to displaying high triplet energies. Time-of-flight mobility measurements revealed good electron mobilities for these compounds ($>10^{-4}$ cm^{2}V^{-1}s^{-1}); following the order **VIIIb** > **VIIIc** > **VIIIa**. Electrophosphorescent device incorporating **VIIIa** as the host, doped with (PPy)$_2$Ir(acac) and 1,3,5-tris(N-phenylbenzimidizol-2-yl)benzene as the electron-transporting material, achieved a high external quantum efficiency of 17.5% and a power efficiency of 59.0 lm W^{-1}. For the same device configuration, **VIIIb**-based device provided values of external quantum efficiency and power efficiency of 14.4% and 50.6 lm W^{-1}, respectively, and **VIIIc**–based device provided values of 5.1% and 12.3 lm W^{-1}, respectively. Superior performance of **VIIIa**-based devices was attributed to the balanced charge recombination, while poor efficiency of **VIIIc**-based device was explained by relatively low triplet energy. Glass-forming

2,4,6-tris[4-(1-naphthyl)phenyl]-1,3,5-triazine with electron drift mobility of 8×10^{-4} cm^2 V^{-1} s^{-1} was also reported (Zeng et al., 2009).

Non-conjugated bipolar compound **IXa** comprising triazine core and bicarbazolyl arms (Fig. 3.9) was synthesized and used for the fabrication of phosphorescent OLEDs (Zeng et al., 2009). The flexible linkages connecting two charge-transporting moieties in the non-conjugated bipolar hybrid molecule increase the entropy because of the more abundant conformations favoring the solubility in organic solvents to facilitate purification and solution processing of the compound. Furthermore, the increased entropy with flexible linkages delivered higher free energy barrier to crystallization from a glassy state, thereby improving morphological stability of the glass as compared to that of relatively rigid conjugated and non-conjugated bipolar hybrid molecules without flexible linkages. Because of the absence of π-conjugation between the two charge-carrier moieties;, that is, triazine and fluorene, the LUMO/HOMO levels and the triplet energies of the two moieties as independent entities retained in the non-conjugated bipolar compound **IXa**. The device with **IXa** as the host of the emitting laser possessed the current density of 0.5 mA/cm^2 and a luminance of 160 cd/m^2, corresponding to current efficiency of 32 cd/A, and external quantum efficiency of 9.2%. The efficiency achieved with **IXa** as the host

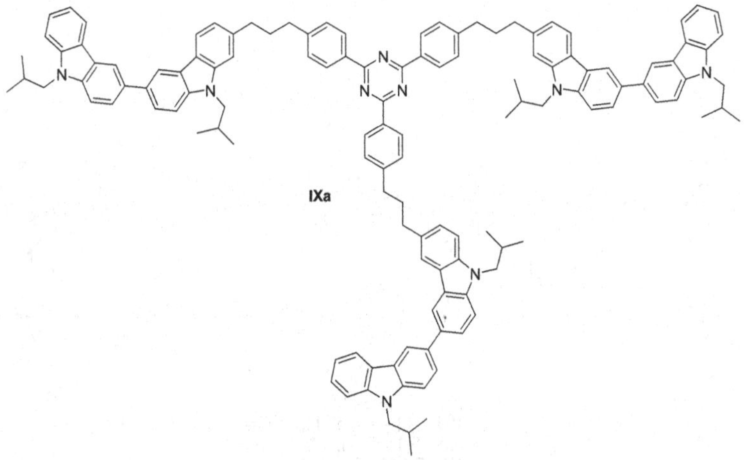

IXa

FIGURE 3.9   IXa.

was reported to be the best among the solution-processed phosphorescent OLEDs prepared using bipolar hosts (Zeng et al., 2009).

Rothmann et al. (2010) reported a series of donor-substituted 1,3,5-triazine derivatives **Xa-f** prepared by nucleophilic substitution of cyanuric chloride with carbazole, 3-methylcarbazole, phenol, and 3,5-dimethylphenol (Fig. 3.10). Symmetric 2,4,6-triscarbazolyl-1,3,5-triazine reported by Inomata et al. (2004) was found to be highly crystalline with a melting point of 465°C. The substitution of the s-triazine core with three 2-methylcarbazole units or three 3-methylcarbazole moieties still yielded materials with a high tendency towards crystallization. Strategic preparation of compounds with either combination of two methylcarbazolyl substituents and one carbazole unit or three different methylcarbazolyl units resulted in materials with slightly decreased crystallization tendencies. Hence, the structural manipulation using stoichiometric mixture of 2-methylcarbazole and 3-methylcarbazole (1:1) as substituents was found to be the most effective way to obtain a less crystalline material **Xa**. The melting temperatures of **Xa-f** ranged from 147 to 335°C. Some of the derivatives formed glasses and the T_gs ranged from 80 to 170°C. The triplet energies of the triazines **XXa-f** were found to be in the range of 2.86–2.96 eV. A maximum brightness of 6900 cd/m^2 and a maximum external quantum efficiency of 10.2% achieved when these materials were used as hosts for blue phosphorescent OLEDs.

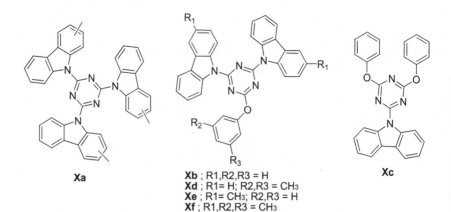

Xa

Xb ; R1,R2,R3 = H
Xd ; R1= H; R2,R3 = CH₃
Xe ; R1= CH₃; R2,R3 = H
Xf ; R1,R2,R3 = CH₃

Xc

FIGURE 3.10   Xa-f.

Donor-acceptor molecules consisting of an electron-deficient triazine core with three fluorene arms substituted with diarylamino **XIa-c** or carbazolyl **XId** electron donors (Fig. 3.11) were reported (Omer et al., 2010). Cyclic voltametry of **XIa-c** showed that the reduction consisted of one wave of single electron transfer to the core, while the oxidation exhibited a single peak of three sequential electron transfer processes with the formation of a trication. Carbazole-containing molecule **XId** after oxidation underwent a subsequent rapid chemical reaction to produce a dimer (*via* the overall coupling of two radical cations with the loss of two protons). The dimer was found to be relatively more easily electro-oxidized than the monomer **XId**. Because of the presence of the acceptor (triazine) center and strong donors in the arms (diarylamines), **XIa-c** exhibited large solvatochromic effects with emissions ranging from deep blue (428 nm) to orange-red (575 nm) depending on the solvent polarity. In dilute solutions, they demonstrated high PL quantum yields of 0.70–0.81. Electro-generated chemiluminescence of **XIa-c** recorded in nonaqueous solutions could even be seen with the naked eye in a well-lit room. Electro-polymerized polymer film of **XId** displayed electrochromic behavior and was colored pale orange in the neutral state and dark green in the oxidized state.

Fluorene substituted triazine derivatives **XIIa-b** (Fig. 3.12) with efficient hole-blocking ability were designed and synthesized (Zhong et al., 2011). Absorption spectrum of **XIIb** was found to be red-shifted by *ca.* 20 nm, compared with 350 nm of **XIIa**. An intramolecular charge-transferred electronic state in **XIIb** caused by the interaction between

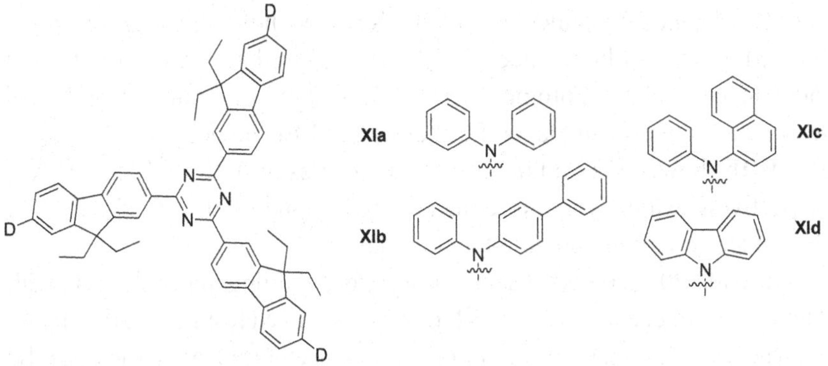

FIGURE 3.11   XIa-d.

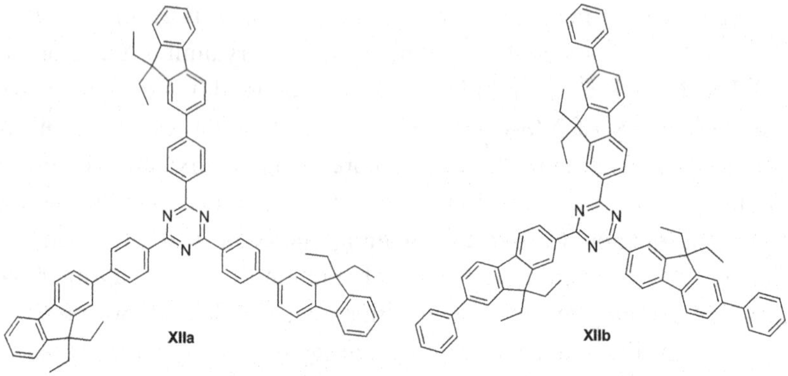

FIGURE 3.12   XIIa-b.

electron-donating fluorene moiety and electron-accepting triazine moiety was found to be well pronounced. Phenyl spacer between fluorene moiety and triazine ring in **XIIa** weakened the charge transfer interaction and as a result, the absorption peak was found to be shifted to shorter wavelengths. The electroluminescent devices based on **XIIa-b** with the configuration of ITO/NPB/Alq$_3$/compounds containing triazine units/LiF/Al showed high luminance of 13000 cd m^{-2} and 18000 cd m^{-2}, respectively, for Alq$_3$/**XIIa** and for Alq$_3$/**XIIb**. The maximum current efficiencies of the devices containing **XIIa** and **XIIb** were found to be remarkably improved from 3.1 cdA^{-1} of the comparative device with a configuration of ITO/NPB/Alq$_3$/Alq$_3$/LiF/Al to 4.0 and 4.9 cd A^{-1}, respectively. The presence of **XIIa** and **XIIb** in the device structures enabled the improvement of balance between hole- and electron-mobility. Polymer light emitting diodes (PLED) containing triazine molecules **XIIa** and **XIIb** as hole-blocking materials showed maximum luminance of 2327 and 2871 cd m^{-2}, respectively, and the values found to be promoted from 1358 cd m^{-2} of comparative device without a hole-blocking layer. Furthermore, PLED devices based on **XIIa** and **XIIb** showed maximum current efficiencies of 0.97, and 1.37 cd A^{-1}, respectively, which were more than four times higher than 0.25 cd A^{-1} of the device containing no hole-blocking layer.

Qu et al. (2007) reported a red–orange light emitting material **XIIIa** with a branched molecular structure. **XIIIa** was assembled from s-triazine, acting as an acceptor group, vinylene conjugated connectors and N-ethylcarbazolyl groups, as donor peripheral groups, by alkaline condensation reaction

(Fig. 3.13). Absorption maximum of **XIIIa** in dilute solution was found to be at 394 nm, while the absorption maximum of the film occurred at 403 nm with a shoulder peak at 503 nm. PL spectrum of **XIIIa** in dilute solution was found to peak at 474 nm. However, the PL spectrum of **XIIIa** film showed two peaks at 483 and 569 nm, which were assigned to the monomer and excimer emissions. **XIIIa** also showed two-photon absorption and two-photon excited fluorescence. EL devices having configuration of ITO/poly(N-vinylcarbazole):**XIIIa**(56 nm)/TPD(5 nm)/Liq(30 nm)/Mg:Ag exhibited good quality white emission.

Multi-branched two-photon absorption triazine derivatives **XIVa-d** (Fig. 3.14) with different donor strength, conjugation length, and direction of charge transfer were reported (Jiang et al., 2011). One-photon absorption and emission spectra revealed that the Stokes shifts of **XIVa-d** increased with an increasing extension of the π-systems and increase of electron-donating ability of the donors, whereas the fluorescence quantum yield exhibited large increase with the introduction of electron-accepting perfluoroalkyl as side groups to the end donor. Investigation of the non-linear absorption properties of these compounds revealed that their two photon absorption cross section values increased with increasing electron-donating strength of the end group. In conjunction with high fluorescence quantum yield, the multi-branched triazine derivative with perfluoroalkyl moiety could find application in two-photon fluorescence bioimaging.

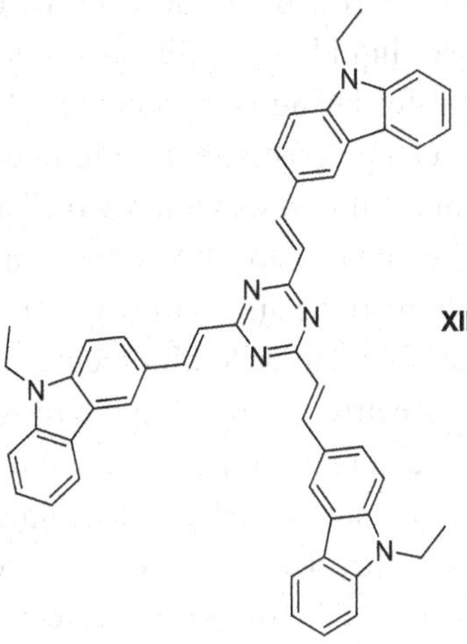

XIIIa

FIGURE 3.13   XIIIa.

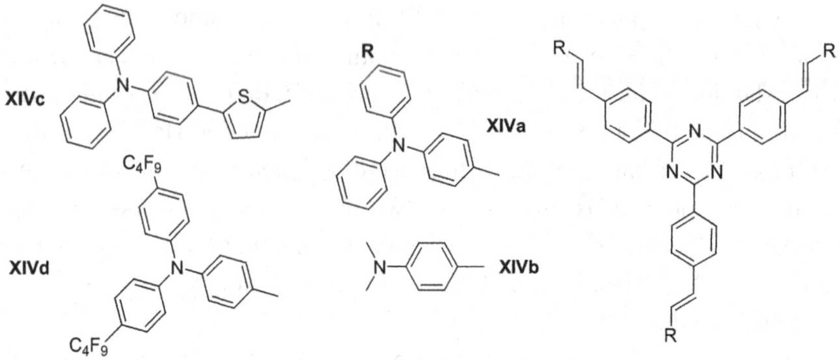

FIGURE 3.14 XIVa-d.

Interesting optical properties exhibited by **XIVa-d** made them potential candidates for the photonics field (Jiang et al., 2011).

Triazine moieties as an electron-transporting central core, separated from thiophene, furan, or EDOT moieties by the p-phenylene spacers were synthesized. The Stiile coupling reaction of bromo substituted triazine derivative yielded **XVa-e** (Fig. 3.15). **XVa-e** displayed excellent redox stability. Dilute solutions of **XVa-e** exhibited fluorescence in the range from 395 to 465 nm with the quantum yield being in the range of 36 to 42%.

Beltran et al. (2010) synthesized tris(triazolyl)triazine core-based compounds **XVIa-c** by applying "click" reactions [64] of the aromatic alkyne and azide precursors (Fig. 3.16). All these 2,4,6-tris(triazolyl)-1,3,5- triazine derivatives formed liquid crystalline phases. **XVIb-c** displayed textures that were typical for hexagonal columnar phases (Col$_h$). Though **XVIa** did not show any characteristic texture in pure state, a Col$_h$ mesophase was observed in miscibility tests with **XVIb**. The maximum absorption wavelengths were located in the UV region, at *ca.* 300 nm. They were attributed to π–π* transitions due to the high absorption coefficients. Compounds **XVIa-c** were found to be luminescent and emitted in the blue-green part of the visible spectrum. A remarkable red-shift in the emission maximum was noted as the number of peripheral alkoxy chains increased. The thin films of the compounds also showed luminescence at room temperature with the emission wavelengths similar to those of the solutions. Electrochemical measurements confirmed the electron-deficient nature of this family of compounds and their potential for electron-transport. **XVIa**

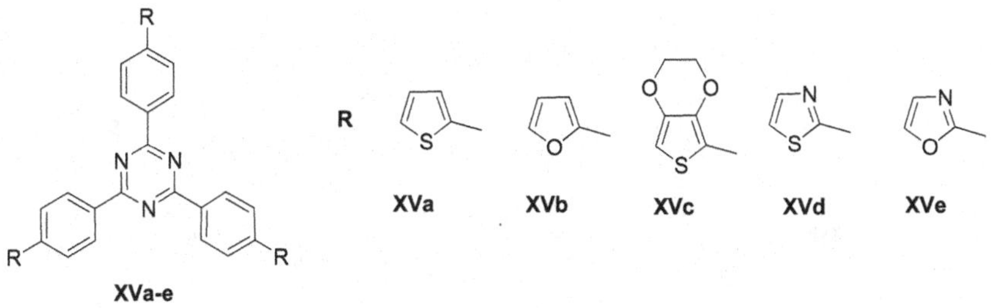

FIGURE 3.15   XVa-e.

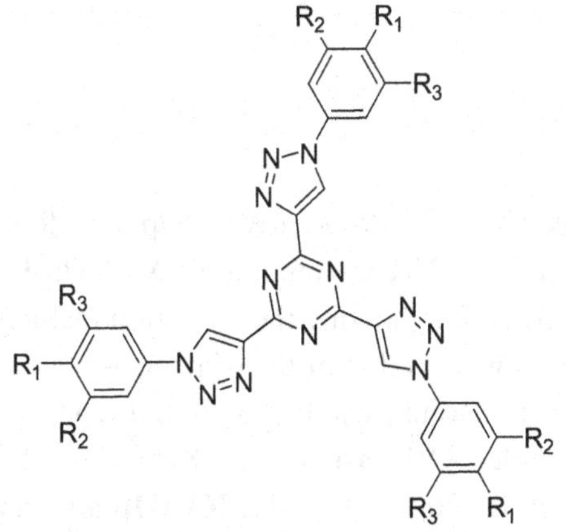

XVIa ; R$_1$ = OC$_{10}$H$_{21}$, R$_2$ =R$_3$ =H
XVIb ; R$_1$ = R$_2$ = OC$_{10}$H$_{21}$, R$_3$ =H
XVIc ; R$_1$ = R$_2$ = R$_3$ = OC$_{10}$H$_{21}$

FIGURE 3.16   XVIa-c.

showed a reversible reduction process with a half-wave potential of –1.28 V. In addition, an irreversible reduction process was found at –2.24 V. **XVIb** also showed the similar behavior, with reversible and irreversible reduction processes (at –1.33 and –2.15 V, respectively). In contrast, only one reversible reduction process was observed, at a half-wave potential of –1.33 V, for **XVIc**.

Tris(tetrathiafulvalene)- and tris(ferrocene)-1,3,5-triazines **XVIIa-b** (Fig. 3.17) with redox properties were reported (Garcia et al., 2009). In the absorption spectra of dilute solutions, the broad band in the visible region at 484 nm for **XVIIa** and 461 nm for **XVIIb** evidenced the occurrence

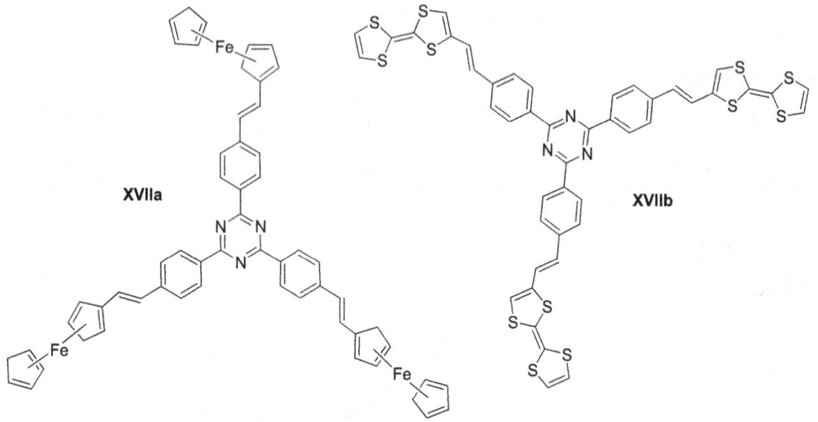

XVIIa
XVIIb
Fe
S
N

FIGURE 3.17 XVIIa-b.

of a donor-acceptor ICT. HOMO-LUMO gap was found to be smaller for **XVIIa** compared to **XVIIb**. Compounds **XXVIIa-b** demonstrated an amphoteric redox behavior showing the oxidation features of the electron-donor substituents as well as that of the triazine as the acceptor unit.

By employing palladium-catalyzed 3-fold coupling methodology, π-conjugated molecules **XVIIIa-e** (Fig. 3.18) based on 1,3,5-triazine were reported (Yasuda et al., 2009). **XVIIIa, XVIIIb** and **XVIIIe** containing flexible alkyl chains exhibited columnar liquid crystalline phases from

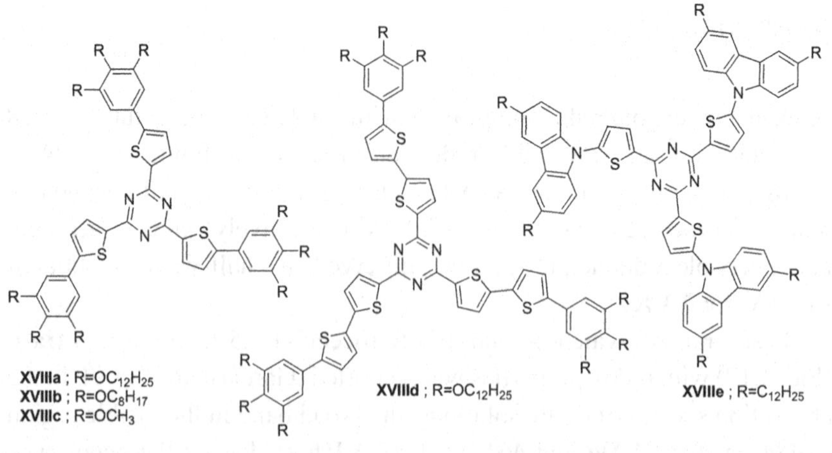

XVIIIa ; R=OC12H25
XVIIIb ; R=OC8H17
XVIIIc ; R=OCH3
XVIIId ; R=OC12H25
XVIIIe ; R=C12H25

FIGURE 3.18 XVIIIa-e.

room temperature upon heating, while the columnar phase of **XVIIId** with an enlarged π-conjugated core existed over relatively higher temperatures in the range of 48–125°C. The time-of-flight experiments revealed that these materials were capable of transporting both holes and electrons in the columnar phases by functioning as ambipolar one-dimensional conducting material. The hole and electron mobilities of **XVIIIa** were estimated to be 3×10^{-5} and 4×10^{-3} cm^2 V^{-1} s^{-1}, respectively, at 100°C. **XVIIIb** also exhibited ambipolar conduction behavior, giving rise to hole and electron mobilities in the order of 10^{-5} and 10^{-3} cm^2 V^{-1} s^{-1}, respectively, in the Col$_h$ phase. Isotropic phase of **XVIIIa** showed lower charge carrier mobility (10^{-6} cm^2 V^{-1} s^{-1}). The hole and electron mobilities of **XVIIIa** were found to be discontinuously increased by *ca.* 1 and 3 orders of magnitude, respectively, at the isotropic Col$_h$ phase transition upon cooling that reflected the formation of ordered one dimensional π-stacked structure in the Col$_h$ phase. The mobility of electrons in the Col$_h$ phases of **XVIIIa-b** was found to be more than 100-times higher than that of holes. This behavior could be originated from the octupolar structure of the propeller-shaped molecules containing both the electron-accepting triazine core and the trigonally ramified electron-donating phenylthiophene units. Hole-mobilities of both **XVIIId** and **XVIIIe** in the Col phases were found to be higher than those of **XVIIIa-b** and reached the values of 1×10^{-3} cm^2 V^{-1} s^{-1}. The enhanced hole-mobilities in **XVIIId-e** as compared to **XVIIIa-b** reportedly arised from an increased intermolecular π-overlap attributing to the expanded electron-donating segments (i.e., the phenylbithiophene units in **XVIIId** and the carbazolylthiophene units in **XVIIIe**) within the columns. The lowest energy absorption maxima of **XVIIIa**, **XVIIId**, and **XVIIIe** were observed at 385, 440, and 415 nm, respectively, in the Col phases. PL emission color of the octupolar materials **XVIIIa-e** was shifted from blue-green to orange by changing the electron-donating segments. **XVIIIe** in the Col phase exhibited an emission maximum at 490 nm, which appeared at a higher energy with a smaller Stokes shift than those of **XVIIIa** (λ_{em} = 519 nm) and **XVIIId** (λ_{em} = 585 nm). The redox properties of **XVIIIa-e** demonstrated that all the compounds underwent both electrochemical oxidation and reduction processes because of their D-A hybrid characteristics.

Our research group reported triazine derived organic materials **XIXa-b** (Fig. 3.19) possessing carbazole or phenothiazine moieties at

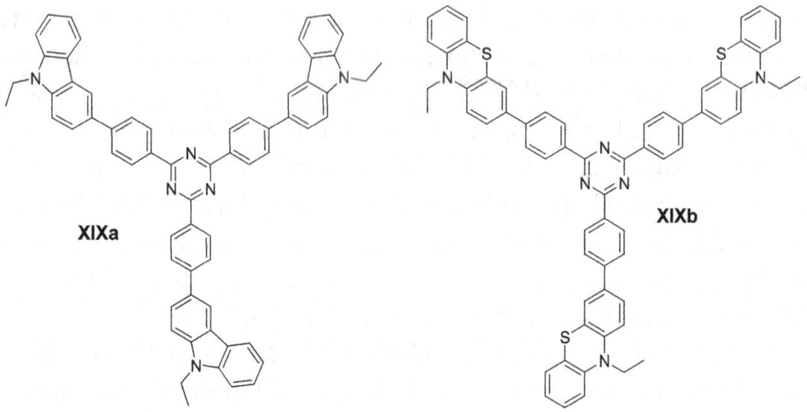

FIGURE 3.19   XIXa-b.

the periphery (Reghu et al., 2013). **XIXa-b** formed molecular glasses. They exhibited high fluorescence quantum yields in dilute liquid and solid solutions. Optical studies performed in different polarity media of **XIXa-b** indicated the occurrence of excited state twisting followed by intramolecular charge transfer. Mono-exponential fluorescence transients were observed for the liquid solutions; however, this character was found to be slightly deviated due to the formation of differently twisted conformers in solid solutions. Non-exponential transients of the neat films were accompanied by the considerable drop in fluorescence quantum yield, which was attributed to exciton-migration-induced quenching at non-radiative decay sites. The redox behavior of **XIXa-b** was influenced by the nature of the attached donor-substituents to the triazine core.

3.5 CONCLUSION

Dendritic organic electroactive compounds reviewed herein demonstrate a wide range of appealing properties like effective charge-transport, operative electroluminescence or high thermal stability. A number of dendritic derivatives showed capability to form glasses with good morphological stability, whereas the compounds containing rigid functional core as well as flexible alkyl surface groups formed organic liquid crystals. Various core-arm combinations in dendritic architecture represent an interesting methodology for the tunabitity of physical properties, in particular, charge-transporting and/or luminescent properties. These compounds can be readily synthesized by applying the so-called "building block approach" together with the use of diversity in cross-coupling reactions. Pyrene-based compounds represented a class of efficient *p*-type organic semiconductors; however, triazine derivatives showed efficient electron-transport or ambipolar-type charge transport in devices. Furthermore, they demonstrated competent redox as well as luminescence characteristics, which made them promising materials for the employment as active layer(s) in explicit devices. Attachment of electron-donor chromophores to the electron-deficient triazine core enables to obtain ambipolar electroactive materials with the opportunity of fine-tuning of decisive properties for device applications. Optoelectronic properties of these dendritic materials are greatly influenced by both the nature and the number of attached substituents at the rim.

ACKNOWLEDGMENT

This research was funded by the European Social Fund under the Global Grant measure.

KEYWORDS

- charge-transport
- dendrimer
- luminescence
- organic semiconductor
- pyrene
- triazine

REFERENCES

1. Abrahams, B. F.; Batten, S. R.; Hamit, H.; Hoskins, B. F.; Robson, R. A Wellsian 'Three-Dimensional' Racemate: Eight Interpenetrating, Enantiomorphic (10, 3)-A Nets, Four Right- and Four Left-Handed. Chem.Comm. 1313–1314 (1996).

2. Anant, P.; Lucasb, N. T.; Ball, J. M.; Anthopoulosc, T. D.; Jacob, J. Synthesis and Characterization of Pyrene-Centered Oligothiophenes. *Synth. Met.* 160, 1987–1993 (2010).

3. Ashizawa, M.; Yamada, K.; Fukaya, A.; Kato, R.; Hara, K.; Takeya, J. Effect of Molecular Packing on Field-Effect Performance of Single Crystals of Thienyl-Substituted Pyrenes. *Chem. Mat.* 20, 4883–4890 (2008).

4. Astruc, D.; Boisselier, E.; Ornelas, C. Dendrimers Designed for Functions: From Physical, Photophysical, and Supramolecular Properties to Applications in Sensing, Catalysis, Molecular Electronics, Photonics, and Nanomedicine *Chem. Rev.* 110, 1857–1959 (2010).

5. Beltran, E.; Serrano, J. L.; Sierra, T.; Gimenez, R. Tris(Triazolyl)Triazine *via* Click-Chemistry: A C_3 Electron-Deficient Core with Liquid Crystalline and Luminescent Properties. *Org. Let.* 12, 1404–1407 (2010).

6. Bernhardt, S.; Kastler, M.; Enkelmann, V.; Baumgarten, M.; Mullen, K. Pyrene as Chromophore and Electrophore: Encapsulation in a Rigid Polyphenylene Shell *Chem: A Eur. J.* 12, 6117–6128 (2006).

7. Brown, A. R.; Pomp, A.; Hart C. M.; de Leeuw D. M. Logic Gates Made from Polymer Transistors and Their Use in Ring Oscillators. *Science* 270, 972–974 (1995).

8. Bundgaard, E.; Krebs, F. C. Low Band Gap Polymers for Organic Photovoltaics. *Sol. Energy Mat. Sol. Cells*, 91, 954–985 (2007).

9. Ceroni, P.; Bergamini, G.; Marchioni, F.; Balzani, V. Luminescence as a Tool to Investigate Dendrimer Properties. *Prog. Polym. Sci.* 30, 453–473 (2005).

10. Chen, H-F.; Yang, S-J.; Tsai, Z-H.; Hung, W-Y.; Wang, T-C.; Wong, K-T. 1,3,5-Triazine Derivatives as New Electron Transport–Type Host Materials for Highly Efficient Green Phosphorescent OLEDs. J. Mat. Chem. 19, 8112–8118 (2009).

10. Crone, B.; Dodabalapur, A.; Gelperin, A.; Torsi, L.; Katz, H. E.; Lovinger, A. J.; Bao, Z. Electronic Sensing of Vapors with Organic Transistors. *Appl. Phys. Let.* 78, 2229–2231 (2001).

12. Crone, B.; Dodabalapur, A.; Lin, Y. Y.; Filas, R. W.; Bao, Z.; LaDuca, A.; Sarpeshkar, R.; Katz, H. E.; Li W. Large-Scale Complementary Integrated Circuits Based on Organic Transistors. Nature 403, 521–523 (2000).

13. De Schryver F. C.; Vosch, T, ; Cotlet, M.; van der Auweraer, M.; Mullen, K.; Hofkens, J. Energy Dissipation in Multichromophoric Single Dendrimers. *Acc. Chem. Res.* 38, 514–522 (2005).

14. Devadoss, C.; Bharathi, P.; Moore, J. S. Energy Transfer in Dendritic Macromolecules: Molecular Size Effects and the Role of an Energy Gradient. *J. Am. Chem. Soc.* 118, 9635–9644 (1996).

15. Garcia, A.; Insuasty, B.; Herranz, M.; Martinez-Alvarez, R.; Martin, N. New Building Block for C_3 Symmetry Molecules: Synthesis of S-Triazine-Based Redox Active Chromophores. *Org. Let.* 11, 5398–5401 (2009).

16. Gingras, M.; Placide, V.; Raimundo, J-M.; Bergamini, G.; Ceroni, P.; Balzani, V. Polysulfurated Pyrene-Cored Dendrimers: Luminescent and Electrochromic Properties. *Chem: A Eur. J.* 14, 10357–10363 (2008).

17. Grazulevicius, J.V.; Strohriegl, P.; Pielichowski, J.; Pielichowski, K. Carbazole-Containing Polymers: Synthesis, Properties and Applications. *Prog. Polym. Sci.* 28, 1297–1353 (2003).

18. Hawker, C. J.; Fréchet, J. M. J. Preparation of Polymers with Controlled Molecular Architecture. A New Convergent Approach to Dendritic Macromolecules. *J. Am. Chem. Soc.* 112, 7638–7647 (1990).

19. Hu, J.; Era, M.; Elsegood, M. R. J.; Yamato, T. Synthesis and Photophysical Properties of Pyrene-Based Light-Emitting Monomers: Highly Pure-Blue-Fluorescent, Cruciform-Shaped Architectures. *Eur. J. Org. Chem.* 72–79 (2010).

20. Huang, J.; Li, G.; Wu, E.; Xu, Q.; Yang, Y. Achieving High-Efficiency Polymer White-Light-Emitting Devices. *Adv. Mat.* 18, 114–117 (2006).

21. Idzik, K. R.; Rapta, P.; Cywinskie, P. J.; Beckert, R.; Dunsch, L. Synthesis and Electrochemical Characterization of New Optoelectronic Materials Based on Conjugated Donor–Acceptor System Containing Oligo-Tri(Heteroaryl)-1,3,5-Triazines. *Electrochim. Acta* 55, 4858–4864, 2010.

22. Inomata, H.; Goushi, K.; Masuko, T.; Konno, T.; Imai, T.; Sasabe, H.; Brown, J. J.; Adachi, C. High-Efficiency Organic Electrophosphorescent Diodes Using 1,3,5-Triazine Electron Transport Materials. *Chem. Mat.* 16, 1285–1291 (2004).

23. Ishi, T.; Yaguma, K.; Thiemann, T.; Yashima, M.; Ueno, K.; Mataka, S. High Electron Drift Mobility in an Amorphous Film of 2,4,6, -Tris[4-(1-Naphthyl)Phenyl]-1,3,5-Triazine. *Chem. Let.* 33, 1244–1245 (2004).

24. Jiang, Y.; Wang, Y.; Wang, B.; Yang, J.; He, N.; Qian. S.; Hua, J. Synthesis, Two-Photon Absorption and Optical Limiting Properties of Multi-Branched Styryl Derivatives Based on 1,3,5-Triazine. *Chem:An Asian J.* 6, 157–165 (2011).

25. Kolb, H. C.; Finn, M. G.; Sharpless, K. B. Click Chemistry: Diverse Chemical Function from a Few Good Reactions. *Angew. Chem. Int. Ed.* 40, 2004–2021 (2001).

26. Krotkus, S.; Kazlauskas, K.; Miasojedovas, A.; Gruodis, A.; Tomkeviciene, A.; Grazulevicius, J. V.; Jursenas, S. Pyrenyl-Functionalized Fluorene and Carbazole Derivatives as Blue Light Emitters. S. *J. Phys. Chem. C* 116, 7561–7572 (2012).

27. Kukuta, A. Infrared Absorbing Dyes. Chapter 12; Plenum Press; New York (1990).

28. Kulkarni, A. P.; Kong, X.; Jenekhe, S. A. High-Performance Organic Light-Emitting Diodes Based on Intramolecular Charge-Transfer Emission from Donor–Acceptor Molecules: Significance of Electron- Donor Strength and Molecular Geometry. *Adv. Funct. Mat.* 16, 1057–1066 (2006).

29. Leu, C-M.; Shu, C. F.; teng, C-F.; Shiea, J. Dendritic Poly(etherimide)s: Synthesis, Characterization, and Modification. *Polymer* 42, 2339–2348 (2001).

30. Liu, F.; Lai, W-Y.; Tang, C.; Wu, H-B.; Chen, Q-Q.; Peng, B.; Wei, W.; Huang, W.; Cao, Y. Synthesis and Characterization of Pyrene-Centered Starburst Oligofluorenes. *Macromol. Rap. Com* 29, 659–664 (2008).

31. Liu, F.; Zou, J-H.; He, Q-Y.; Tang, C.; Xie, L-H.; Peng, B.; Wei, W.; Cao, Y.; Huang, W. Carbazole End-Capped Pyrene Starburst with Enhanced Electrochemical Stability and Device Performance. *J. Poly. Sci. Part A: Poly. Chem.* 48, 4943–4949 (2010).

32. Lygaitis, R.; Getautisc, V.; Grazulevicius, J. V. Hole-Transporting Hydrazones. *Chem. Soc. Rev.* 37, 770–788 (2008).
33. Maes, W.; Dehaen, W. Synthetic Aspects of Porphyrin Dendrimers. *Eur. J. Org. Chem.* 4719–4752 (2009).
34. Murayama, T. Organic Photoconductive Materials Used for Electrophotographic Organic Photoreceptors. *J. Synth. Org. Chem. Japan.* 57, 541–551 (1999).
35. Omer, K. M.; Ku, S-Y.; Chen, Y-C.; Wong, K-T.; Bard, A. J. Electrochemical Behavior and Electrogenerated Chemiluminescence of Star-Shaped D−A Compounds with a 1,3,5-Triazine Core and Substituted Fluorene Arms. *J. Am. Chem. Soc.* 132, 10944–10952 (2010).
36. Operamolla, A.; Farinola, G. M. Molecular and Supramolecular Architectures of Organic Semiconductors for Field-Effect Transistor Devices and Sensors: A Synthetic Chemical Perspective. *Eur. J. Org. Chem.* 423–450 (2011).
37. Pizzoferrato, R.; Ziller, T.; Micozzi, A.; Ricci, A.; Sterzo, C. L.; Ustione, A.; Oliva, C.; Cricenti, A. Suppression of the Excimer Photoluminescence in a Poly(Arylene–Ethynylene) *co*-Polymer. *Chem. Phys. Let.* 414, 234–238, 2005.
38. Puntoriero, F.; Ceroni, P.; Balzani, V.; Bergamini, G.; Vogtle, F. Photoswitchable Dendritic Hosts: A Dendrimer with Peripheral Azobenzene Groups. *J. Am. Chem. Soc.* 129, 10714–10719 (2007).
39. Qu, B.; Chen, Z.; Xu, F.; Cao, H.; Lan, Z.; Wang, Z.; Gong, Q. Colour Stable White Organic Light-Emitting Diode Based on a Novel Triazine Derivative. *Org. Electron.* 8, 529–534 (2007).
40. Reghu, R. R.; Grazulevicius, J. V.; Simokaitiene, J.; Matulaitis, T.; Miasojedovas, A.; Kazlauskas, K.; Jursenas, S.; Data, P.; Lapkowski, M; Zassowski, P. Glass Forming Donor-Substituted S-Triazines: Photophysical and Electrochemical Properties. *Dyes Pigm.* 97, 412–422 (2013).
41. Reghu, R. R.; Grazulevicius, J. V.; Simokaitiene, J.; Miasojedovas, A.; Kazlauskas, K.; Jursenas, S.; Data, P.; Karon, K.; Lapkowski, M.; Gaidelis, V.; Jankauskas, V. Glass-Forming Carbazolyl- and Phenothiazinyl- Tetra Substituted Pyrene Derivatives: Photophysical, Electrochemical and Photoelectrical Properties. *J. Phys. Chem. C* 116, 15878–15887 (2012).
42. Rothmann, M. M.; Haneder, S.; Como, E. D.; Lennartz, C.; Schildknecht, C.; Strohriegl, P. Donor-Substituted 1,3,5-Triazines as Host Materials for Blue Phosphorescent Organic Light-Emitting Diodes. *Chem. Mat.* 22, 2403–2410 (2010).
43. Sadler, K.; Tam, J. P. Peptide Dendrimers: Applications and Synthesis. *Rev. Mol. Biotechnol.* 90, 195–229 (2002).
44. Someya, T.; Katz, H. E.; Gelperin, A.; Lovinger, A. J.; Dodabalapur, A. Vapor Sensing with α, ω-Dihexylquarterthiophene Field-Effect Transistors: The Role of Grain Boundaries. *Appl. Phys. Let.* 81, 3079–3081 (2002).
45. Sonar, P.; Soh, M. S.; Cheng, Y. H.; Henssler, J. T.; Sellinger, A. 1,3,6,8-Tetrasubstituted Pyrenes: Solution-Processable Materials for Application in Organic Electronics. *Org. Let.* 12, 3292–3295 (2010).
46. Sonogashira, K.; Tohda, Y.; Hagihara, N. A Convenient Synthesis of Acetylenes: Catalytic Substitutions of Acetylenic Hydrogen with Bromoalkenes, Iodoarenes and Bromopyridines. *Tetrahed. Let.* 16, 4467–4470 (1975).

47. Strohriegl, P.; Grazulevicius, J. V. Charge-Transporting Molecular Glasses. *Adv. Mat.* 14, 1439–1452 (2002).
48. Thomas, J.; Christenson, C. W.; Blanche, P. –A.; Yamamoto, M.; Norwood, R. A.; Peyghambarian, N. Photoconducting Polymers for Photorefractive 3D Display Applications. *Chem. Mat.* 23, 416–429 (2011).
49. Turnbull, W. B.; Stoddart, J. F. Design and Synthesis of Glycodendrimers. *Rev. Mol. Biotechnol.* 90, 231–255 (2002).
50. Venkataramana, G.; Sankararaman, S. Synthesis, Absorption, and Fluorescence-Emission Properties of 1,3,6,8-Tetraethynylpyrene and Its Derivatives. *Eur. J. Org. Chem.* 4162–4166 (2005).
51. Vogtle, F.; Richardt, G.; Werner, N. Dendrimer Chemistry Concepts, Syntheses, Properties, Applications. Wiley; Weinheim, Germany (2009).
52. Walker, B.; Kim, C.; Nguyen, T.-Q. Small Molecule Solution-Processed Bulk Heterojunction Solar Cells. *Chem. Mat.* 23, 470–482 (2011).
53. Walter M. V.; Malkoch, M. *Chem. Soc. Rev.* 41, 4593–4609 (2012).
54. Wan, Y.; Yan, L.; Zhao, Z.; Ma, X.; Guo, Q.; Jia, M.; Lu, P.; Ramos-Ortiz, G.; Maldonado, J. L.; Rodriguez, M.; Xia, A. Gigantic Two-Photon Absorption Cross Sections and Strong Two-Photon Excited Fluorescence in Pyrene Core Dendrimers with Fluorene/Carbazole as Dendrons and Acetylene as Linkages. *J. Phys. Chem. B*, 114, 11737–11745 (2010).
55. Wang, J-L.; Zhong, C.; Tang, Z-M.; Wu, H.; Ma, Y.; Cao, Y.; Pei J. Solution-Processed Bulk Heterojunction Photovoltaic Cells from Gradient π-Conjugated Thienylene Vinylene Dendrimers. *Chem: An Asian J.* 5, 105–113 (2010).
56. Weder, C.; Wrighton, M. S. Efficient Solid-State Photoluminescence in New Poly(2,5-Dialkoxy-P-Phenyleneethynylene)s. *Macromol.* 29, 5157–5165 (1996).
57. www.lg.com/au/oled-tv
58. www.liternity.com
59. www.lumiblade-experience.com
60. Xia, R.; Lai, W-Y.; Levermore, P. A.; Huang, W.; Bradley, D. C. Low-Threshold Distributed-Feedback Lasers Based on Pyrene-Cored Starburst Molecules with 1,3,6,8-Attached Oligo(9,9-Dialkylfluorene) Arms. *Adv. Funct. Mat.* 19, 2844–2850 (2009).
61. Xu, F.; Wang, Z.; Gong, Q. Synthesis, Characterization, and Optical Properties of Two-Photon-Absorbing Octupolar Molecule with an *S*-Triazine Core. *Opt. Mat.* 29, 723–727 (2007).
62. Yasuda, T.; Shimizu, T.; Liu, F.; Ungar, G.; Kato, T. Electro-Functional Octupolar π-Conjugated Columnar Liquid Crystals. *J. Am. Chem. Soc.* 133, 13437–13444 (2011).
63. Zeng, L.; Lee, T. Y-H.; Merkel, P. B.; Chen, S. H. A New Class of Non-Conjugated Bipolar Hybrid Hosts for Phosphorescent Organic Light-Emitting Diodes. J. Mat. Chem. 19, 8772–8781 (2009).
64. Zerkowski, J. A.; Seto, C. T.; Whitesides, G. M. Solid-State Structures of Rosette and Crinkled Tape Motifs Derived from the Cyanuric Acid Melamine Lattice. *J. Am. Chem. Soc.* 114, 5473–5475 (1992).
65. Zhao, Z.; Li, J-H.; Chen, X.; Wang, X.; Lu, P.; Yang, Y. Solution-Processable Stiff Dendrimers: Synthesis, Photophysics, Film Morphology, and Electroluminescence. *J. Org. Chem.* 74, 383–395 (2009).

66. Zhong, C.; Duan, C.; Huang, F.; Wu, H.; Cao, Y. Materials and Devices toward Fully Solution Processable Organic Light-Emitting Diodes. *Chem. Mat.* 23, 326–340 (2011).
67. Zhong, H.; Lai, H.; Fang, Q. New Conjugated Triazine Based Molecular Materials for Application in Optoelectronic Devices: Design, Synthesis, and Properties. *J. Phys. Chem. C,* 115, 2423–2427 (2011).

CHAPTER 4

RHEOLOGICAL CHARACTERISTICS OF LINEAR LOW DENSITY POLYETHYLENE – FUMED SILICA NANOCOMPOSITES

V. GIRISH CHANDRAN and SACHIN WAIGAONKAR

Department of Mechanical Engineering, BITS Pilani K K Birla Goa Campus, Zuarinagar Goa, India–403726; Email: p2011407@goa.bits-pilani.ac.in; sdw@goa.bits-pilani.ac.in

CONTENTS

4.1 Introduction... 63

4.2 Materials and Methods.. 67

4.3 Results and Discussions.. 70

4.4 Conclusion 74

Authors' Contribution 74

Keywords 75

References.. 75

4.1 INTRODUCTION

4.1.1 RHEOLOGY

The term Rheology introduced by Eugine C Bingham in 1920, is currently used as a semi-quantitative tool in study of polymer melt studying the deformation and flow under stress. The relationship between polymer structure and the measurable properties like viscosity, strain rate and visco-elastic properties are currently investigated by many researchers and remains a challenge to address the properties of commercially available polymers. (Dealy and Larson, 2006; Gahleitner, 2001). During the past two decades, several attempts are made to manufacture precise, reliable, simple to operate and economically attractive in-line and on-line rheometers (Mould et al., 2011). Melt Flow Index (MFI) is one of the most popular rheological measures for specifying the flow properties of polymer melt and its characterization. American Society for Testing and Materials specifies the standard procedure for measurement of melt flow rates for thermoplastics by extrusion (ASTM-D1238). The melt flow index represents the weight of polymer passing through a standard die (Inside bore- 2.095 ± 0.0051 mm) at specified temperature and load in 10 minutes. The MFI of a thermoplastic depends on its molecular structure, degree of polymerization and branching, micro and nano particulate additives, temperature and shear rate. Commercially, a polymer blend procured for processing is specified by mentioning the density and MFI along with other parameters. The MFI measures the flowability of polymer melt and is broadly identified as the reciprocal of dynamic viscosity (η). The viscosity (ratio of shear stress to shear rate) of polymer melt reduces with increased shear rate $\left(\dfrac{dy}{dt} \right)$ for constant temperature as given in Fig. 4.1. This pseudoplastic behavior is typical for thermoplastic polymers.

The dynamic viscosity of polymer is represented as tan α where as tan β represents the apparent viscosity. According to the theory of viscoelasticity, at very low shear rates the viscosity is independent of shear rate, and the limiting value of viscosity is called zero shear viscosity (η_0). This is an important material constant and plays an important role in molecular rheology and serves as an indicator for molecular weight. The direct measurement of zero shear viscosity is often difficult, since the standard rheometers do not provide reliable data at sufficiently low shear rates to reach the region of Newtonian behavior. The zero shear viscosity has strong dependence on molecular weight.

Polyolefines are the most widely used thermoplastic today, and are based on alkalene-1 monomers or α- olefines comprising of ethylene

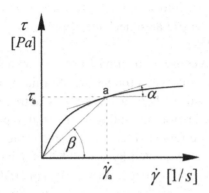

FIGURE 4.1 Viscosity coefficients.

based and propylene-based polymers. Among the ethylene based thermo plastics Linear Low Density Polyethylene (LLDPE) is widely used for many outdoor engineering and domestic applications due to its high strength, impact resistance, environmental stress resistance and broader processing window. Other members in the family include Low Density Polyethylene (LDPE), typically used for making thin films, and High Density Polyethylene (HDPE) typically used for making pipes, bottles and other engineering and structural applications. The thermo plastics, unlike thermosets and metals posses lower mechanical properties like tensile strength, modulus and impact resistance. Various micro scale and nano scale additives are added to polymer matrix to improve mechanical properties, environment resistance, aesthetics etc. The micro scale additives do not fall in the category of reinforcement and hence adversely affect the mechanical properties owing to their hardness, while elastomer like particle may increase toughness but result in reduction of strength. Fumed silica is proving to be a promising additive used in polymers to provide high rigidity, impact toughness and increased tensile strength in Poly ethylene terephthalate, epoxy composites, Poly ethylene naphthalate, Ployamide etc. (Hahm, 2004; Jiang, 2013; Seong, 2003; Yang, 1998). The addition of nano particles with large specific surface area (BET surface area) to polymer matrices lead to amplification of a number of rather distinct molecular processes resulting from interactions between chains and solid surfaces. Another basic characteristic of nano-particles in polymer matrices is the tendency for the particles to associate into extended structures that can dominate the rheological, visco-elastic and mechanical properties of the nano-com- posites (Jancer, 2010).

Different additives are used with LLDPE for enhancing the mechan- ical properties and improved product life. Normally, micro scale addi- tives are blended with LLDPE using a screw extruder and is made in pallets or pulverized before moulding to get desired products. Addition of nano particles like fumed silica, organo clays, metal nano particles, titanium oxide are investigated by researchers for increased mechani- cal properties. In a series of experimental runs of HDPE nano compos- ites containing up to 5 wt. % of FS, it was found that tensile strength increased by increasing silica content up to 2.5 wt. % (Chrissafis, 2009). It was established that the degree of matrix reinforcement gets considerably affected by the extent of matrix crystallinity. As a result of adding FS into the melted matrix, mobility of chains in contact with silica particles became reduced, which caused substantial changes in morphology of these semi crystalline nano composites (Kalfus, 2008)

4.1.2   EFFECT OF FUMED SILICA AS NANO FILLER

Several researches have revealed that the presence of FS in polymer blends results in microstructural changes. Addition of FS suppressed spherulitic growth formation in LDPE during melt mixing, creating well organized lamellar amorphous regions in the crystalline structure leading to enhanced mechanical properties (Panaitescu, 2013). During polymer solidification spherulites develop and grows till it impinges another, forming randomly oriented structures with weak spherulite boundaries. These boundaries are weak zones and tend to decrease the mechanical properties of the polymer. Presence of fumed silica in LLDPE polymer blend, occupies the amorphous regions in the polymer matrix and tend to increase the melt viscosity resulting in reduction of spherulitic growth. The atomic force microscopy studies revealed non- existence of spherulitic growth in PE-FS blends. In the case of FS nano- composites, filler inter-particle interactions dominate the viscoelastic behaviors. As a consequence, the rupture of the filler network is the first-order mechanism in the Payne effect. On the contrary, the polymer – particle interactions, which lead to a strong modification of the chain relaxation process, is the dominant mechanism with colloidal silica at identical filler concentrations. The mechanism of disentanglement and entanglement should be the dominant one in the Payne and thixotropy (modulus recovery) effects with spherical colloidal nano-composites. From a rheological point of view, a direct consequence of incorporation of fillers in molten polymers is a significant change in their steady shear viscosity behavior and the visco-elastic properties. (Cassagnau, 2008, 2013).

Though the addition of FS nano particles in thermoplastics result in increased viscosity, the factors like nucleation affect in certain polymers. In one such study on the nucleation activity, which indicated the influence of the filler on the polymer matrix, revealed that the FS nano particle had a good nucleation effect on Polyethylene Naphthalate (PEN). The ordinarily higher melt viscosity of PEN was reported to decrease on addition of fumed silica (Seong, 2003). Similar decrease in melt viscosity was observed by Im et al. (Im, 2002) in Poly Ethylene Terephthalate (PET) matrix, and Cho et al. (Cho, 2001) reported the reduction of melt viscosity for nylon-6 nano-composites filled with organo-clays, as measured by capillary rheometry. They suggested two possible reasons for the reduction of the melt viscosity, when nano-size fillers were added to the polymer matrix. The first was the slip between the polymer matrix and the filler, and the second was the degradation of the polymer matrix due to the high shear force and heat during melt compounding. However, when silica nano particles were used as the filler, they would act as a lubricant during melt compounding rather than in the second way described above, because it has spherical shapes and smooth nonporous surfaces, which would lower the friction coefficient (Seong, 2003). The effects of silica nano-particles on the phase separation of Poly (Methyl Methacrylate)/Poly (Styrene Acrylonitrile) (PMMA/SAN) blends were studied by Gao et al. (2012) showed small amount of the silica nano-particles in PMMA/SAN blends will significantly change the phase diagram, which is related to the selective location of silica in PMMA (Gao, 2012).

4.1.3   TORQUE RHEOMETRY

The rheological behavior of polymer melt can also be studied by torque rheometry. It measures the viscosity related torque generated by the resistance of a material to the shearing force. The shearing force can be applied by parallel plates, cup and cone, rotating screws etc. Typical analyses include the study of processing behavior, influence of additives, thermal sensitivity, shear sensitivity, compounding behavior, and others. The torque rheometry studies are generally used for analyzing the cross linking of polymers and the effect on torque increase along with the processing temperatures due to the usage of filler particles. Parallel plate torque rheometry studies were used to study the chemical cross linking of Poly Vinyl Chloride (PVC) induced by metallic mercaptides (Ba and Mg salts of 2-dibutylamino-4, 6-dithio-1,3,5-triazine) combined with various thermal stabilizer combinations (calcium/zinc and barium/zinc stearates). The extent of cross-linking was determined by measuring the solvent (tetrahydrofuran) insoluble gel content. The cross linking reaction measured by torque and parallel plate rheometry, showed that the magnesium salt of the 2-dibutylamino-4, 6-dithio-1,3,5-triazine was more effective than the barium salt in cross-linking the PVC (Rosales, 2000). Hale et al. investigated effects of cross-linking reactions and order of mixing on properties of compatibilized Poly butelene terephthalate/ABS blends and its impact on mechanical properties using a single screw extruder. The blend viscosity was much higher than that of the individual components indicating the possibility of chemical reaction. The impact strength of ABS-45/PMMA 30/5 was slightly lower than ABS-45/SAN 30/5 blends, indicating that the presence of an acrylic polymer with no epoxy groups, which is miscible with the SAN matrix, did have a small effect on ABS impact properties (Hale, 1999). Similar studies on PET/Poly carbonate (PC) blends were carried out using elongation rheometry by Robinson et al. Addition of PC at low levels (10–20%) permitted the thermo elastic processing of PET over a wider temperature range, and suggested that uni-axial orientation and significant strain-induced crystallization in the PET -phase can be achieved at lower strain levels (Robinson, 1996). The addition of nano particles in polymer blend is associated with increase in viscosity and hence increased torque is required for polymer processing. With the addition of hydrophobic and hydrophilic FS of different surface area and particle size in solvent based polyurethane additives, a noticeable increase in the viscosity, negative thixotropy and increase in storage modulus was seen (Martinez, 1996). These modifications were pronounced in the adhesives that contained hydrophilic FS.

4.2 MATERIALS AND METHODS

The polymer used in this study was LLDPE of grade R350 A 42 having melt flow index of 4.2 g/10 min and density of 935 Kg/m³ supplied by GAIL India Limited.

Fumed Silica used was Aerosil-200 (BET surface area- 200 m²/g) supplied by Evonik industries.

LLDPE-FS nano-composite blends (1–8 wt. % FS) were prepared by dry mixing and melt mixing using high-speed mixer and lab scale single screw extruder. The powder and pallets produced by blending were dried and stored in airtight containers to reduce the chance of moisture entrapment. The MFI was measured at 190°C/2.16 Kg, for the blends as per ASTM-D1238 using the melt flow indexer from Dynisco. The torque rheometry was carried out at 150°C and 210°C at 60 rpm as referred in ASTM-D2538. Figure 4.2 shows the MFI testing equipment. It is equipped with encoder for setting the flag lengths and manual dead weights to be loaded on top of piston as shown.

Figures 4.3 and 4.4 show the lab scale extruder with torque rheo attachment. The rheo attachment is a twin-screw rheometer and can be attached to the extruder servo motor using a coupling. The net chamber volume of the mixer chamber is 69 cc and 70% of the chamber is filled with polymer blend while doing the testing.

The melt flow indexer is essentially an extrusion plastometer, which consists of a chamber for melting the polymer along with a die and dead weight piston arrangement. The melt chamber is cylindrical in shape and thermostatically controlled to ensure constant set temperature during the testing. The essential features of the melt flow indexer are given in Fig. 4.5. As per ASTM-D1238 the steel cylinder shall be 50.8 mm in diameter, 162 mm in length with a smooth straight hole of 9.5504 ± 0.0076 mm in diameter displaced 4.8 mm from the central axis to accommodate

FIGURE 4.2   Melt flow indexer.

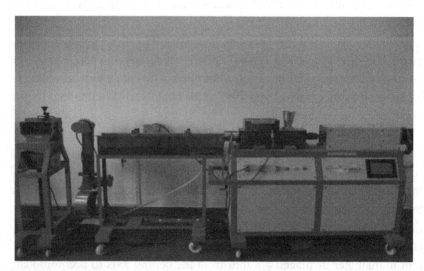

FIGURE 4.3   Lab scale extruder.

FIGURE 4.4   Torque Rheo attachment.

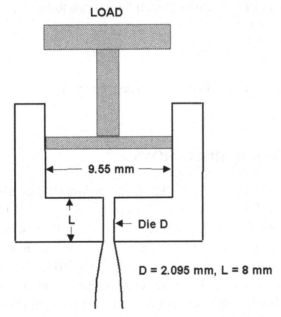

FIGURE 4.5   Schematic of melt flow indexer.

thermal sensors. The piston compresses the polymer melt inside this thermally controlled cylindrical melt chamber, which has a steel die of 2.0955 ± 0.0051 mm at the bottom through which the extrudate flows.

The load is applied on piston by mounting weights directly on piston or by using pneumatic actuation. Melt flow index can be determined by manual (Method A) and auto modes (Method B). In manual operation the extrudate is cut at specified time interval (usually 20 sec) and is weighed in a precision balance to determine the melt flow rate. In the auto mode the weighing of extrudate is avoided by specifying the melt density of the polymer as an input in the melt flow indexer. The melt flow indexer measures the Melt Volume Rate (MVR) as the ratio between the volume of polymer melt flown through the die at constant temperature and pressure to the time taken and can be calculated according to equation 1.

$$MVR\ (cc\ /\ 10min) = V.s\ /\ t. \tag{1}$$

where s is the factor of standard time (10 min, s=600).

The mass (M) of polymer melt can be measured separately using a precision balance for obtaining the MFI according to Eq. (2).

$$MFI\ (g\ /\ 10min) = M.s\ /\ t. \tag{2}$$

The ratio of MFI to MVR thus gives the melt density of polymer extrudate.

4.3   RESULTS AND DISCUSSIONS

The variation of MFI for LLDPE-FS nano-composite is shown in Table 4.1. It can be noticed that addition of FS from 1 to 8 wt. % in LLDPE result in reduction of the MFI. This clearly indicates that the viscosity of polymer melt increases with increased FS composition. This is in accordance with published literature (Panaitescu, 2013; Lau, 2011). The trend of this variation is shown in Fig. 4.6. It was observed that the melt characteristics like MFI, melt viscosity and shear rate did not vary significantly for both blending techniques till 2% FS composition.

TABLE 4.1 Variation of MFI

Blend	MFI (g/10 min)	
	Melt mixing	Dry Mixing
Natural LLDPE	4.49	4.51
LLDPE + 1% FS	4.23	4.16
LLDPE + 2% FS	3.99	3.76
LLDPE + 3% FS	3.75	3.43
LLDPE + 4% FS	3.65	2.96
LLDPE + 5% FS	3.53	2.78
LLDPE + 8% FS	3.14	1.73

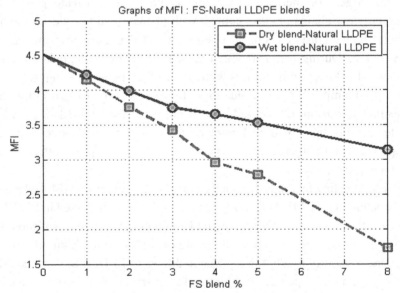

FIGURE 4.6   Variation of MFI (190°C/2.16 Kg) as per ASTM-D1238.

The addition of FS decreased the MFI and hence increased the viscosity of polymer melt. This can be attributed to restricted movement of polymer chains in the melt. The mobility of polymer chains are affected by the addition of FS as it occupies the amorphous region and create hindrance to polymer chain movement due to its comparative size and increased chances of polymer entanglements. More heat input is required to increase the viscosity of LLDPE-FS blends with larger FS content. The additional amount of heat needed to increase viscosity tends to increase significantly after 2% of FS composition. The effect of increased viscosity and heat retention capacity of FS was verified with the increased torque levels and polymer degradation at higher FS contents. The polymer blend with 3% FS degraded at prolonged exposure to heat at 210°C while degradation was observed at 150°C for blend with 8% FS in torque rheometry analysis. The FS having lower heat capacity (740 J/Kg K) than LLDPE (2080 J/Kg K) tend to attain increased temperatures in the polymer matrix for the same heat input. The poor heat dissipation tend to melt the lower molecular weight polymer chains and the side chains of the larger polymer chains in close contact with FS nano particles. The polymer chains in close contact with FS will tend to oxidize due to higher temperature, causing discoloration and degradation at increased FS content. During Torque rheometry, the polymer blend degradation was observed at lower temperatures with larger FS content. This suggests the increased temperature of FS due to prolonged exposure to heat in torque rheometry, causing the degradation of polymer. Thus fumed silica provides rigidity by providing hindrance for polymer chain movements is apparent with the increased melt viscosity but may not prove effective at higher temperature.

The blending methods did not have significant effect on MFI for lower FS content in LLDPE. At lower FS content FS is well dispersed in LLDPE even with dry mixing. With increased FS content the proper dispersion of FS is achieved only through melt mixing or wet blending. Melt mixing is the preferred way to add micro sized additives to LLDPE to enhance the various desired properties. The torque rheometric studies showed a maximum of 30% torque increase with 8% addition of FS in LLDPE both at 150°C and 210°C as shown in Figs. 4.7 and 4.8.

The initial torque increase suggests the melting of polymer blend inside the chamber. The initial torque recorded 60–70 Nm at 150°C and 35–40 Nm at 210°C for the polymer blends tested. After melting the torque generally remained constant indicating the thermal stability of polymer blend. The presence of cross-linking is usually identified by the

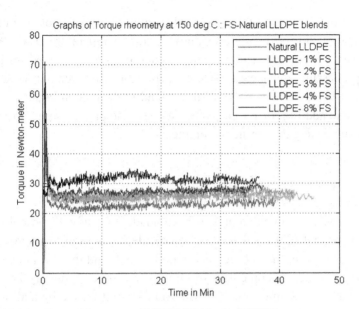

FIGURE 4.7   Torque Rheometry of LLDPE-FS blends @ 150°C.

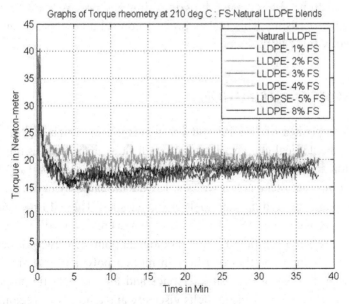

FIGURE 4.8   Torque Rheometry of LLDPE-FS blends @ 210°C.

sharp torque increase after a period of thermal stability of polymer blend. The torque showed increased values at lower temperature (150°C) and increased FS content. The increased torque need to accounted for polymer processing and increasing the temperature for reduction of torque may not be suitable as polymer blend tend to degrade at larger FS contents. The addition of FS may thus decrease the processing temperature while increasing the torque requirements.

4.4   CONCLUSION

In this chapter, an attempt has been made to study the rheological properties of LLDPE-FS blends. Nanoscale FS is used to improve the mechanical properties of the polymers. It was found that MFI decreases with increasing FS content in LLDPE. The torque rheometry has suggested that a maximum of 30% variation in melt torque by addition of FS (max 8 wt. %). The degradation of LLDPE-FS blend was noticed at higher temperatures. The combined effect of adding micro sized additives for environment resistance, coloring, etc., along with nano particles need to be studied for fully controlling the process parameters for polymer processing. This makes process control a key task when FS is blended with LLDPE.

AUTHORS' CONTRIBUTION

Nano-composites of LLDPE and FS blends are investigated, for studying the rheological characteristics and melt flow. FS is used as reinforcement for improving mechanical properties in thermoplastics. LLDPE-FS nano-composite blends (1–8 wt. % FS) were prepared by dry mixing and melt mixing for studying the melt characteristics. The MFI measured at 190°C/2.16 Kg as per ASTM-D1238, shown decrease in MFI values with increased FS composition. From a rheological point of view, a direct consequence of incorporation of fillers in molten polymers is a significant change in their shear viscosity behavior and the viscoelastic properties. The increased polymer blend melts viscosity and amorphous regions may affect the blend preparation with various micro sized additives added for coloring, UV protection and other desired properties. The rheological characteristics of the LLDPE-FS nano composite with various compositions were studied using torque rheometer.

KEYWORDS

- **fumed silica**
- **LLDPE**
- **melt flow index**
- **nano-composites**
- **torque rheometry**

REFERENCES

1. Cassagnau, P. Linear Visco-elasticity and Dynamics of Suspensions and Molten Polymers filled with Nano-particles of Different Aspect Ratios. *Polymer*. (2013).
2. Cassagnau, P. Melt Rheology of Organoclays and Fumed Silica Nanocomposites. *Polymer*. 49, 2183–2196 (2008).
3. Cho, J. W.; Paul, D. R. Nylon 6 Nano Composites by Melt Compounding, *Polymer* 42,1083–1094 (2001).
4. Chrissafis K.; Paraskevopoulos K M.; Pavlidou E.; Bikiaris D. Thermal degradation mechanism of HDPE nanocomposites containing fumed silica nano particles. *Thermochimica Acta*, 485(1–2), 65–71 (2009).
5. Dealy, J. M., Larson R. G. *Structure and Rheology of Molten Polymers: From Structure to Flow Behavior and Back Again*; Hansel: Munich, (2006).
6. Gahleitner, M. Melt rheology of polyolefins. *Prog. Polym. Sci.* 26, 895–944 (2001).
7. Gao, J.; Huang, C.; Wang, N.; Yu W.; Zhou C. Phase Separation of Poly (methyl methacrylate)/Poly (styrene-co-acrylonitrile) Blends in the Presence of Silica Nanoparticles. *Polymer.* 53, 1772–1782 (2012).
8. Hahm, W.G.; Myung, H.S.; Im, S. S. Preparation and Properties of in situ Polymerised Poly(ethylene terephthalate)/Fumed Silica Nanocomposites *Macromolecular research.* 12(1), 85–93 (2004).
9. Hale, W. K.; Paul D. R. Effect of Crosslinking Reactions and Order of Mixing on Properties of Compatibilized PBT/ABS Blends. *Polymer.* 40, 3665–3676 (1999).
10. Im, S. S.; Chung, S. C.; Hahm, W. G.; Oh, S. G. Polyethylene Terephthalate Nano Composites filled with Fumed Silica by Melt Compounding. *Macromol Reserve.* 10 (4), 221–229 (2002).

11. Jancar, J.; Douglas, J. F; Starr, F.W.; Kumar, S.K.; Cassagnau, P.; Lesser, A.J.; Sternstein, S. S.; Buehler, M. J. Feature article, Current Issues in Research on Structure and Property Relationships in Polymer Nanocomposites. *Polymer.* 51, 3321–3343 (2010).

12. Jiang, T.; Kuila, T.; Kim, N. H.; Ku, B. C.; Lee, J. H. Enhanced Mechanical Properties of Silanized Silica nanoparticle Attached Graphene Oxide/Epoxy Composites. *Compos. Sci. Technol.* 79, 115–125 (2013).

13. Kalfus J.; Jancer J.; Kucera J. Effect of weakly interacting nanofiller on the morphology and viscoelastic response of polyolefins, *Polym Eng Sci*, 48(5), 889–894 (2008).

14. Lau, K. Y.; Vaughan, A.S.; Chen, G.; Hosier, I. L. On the Effect of Nanosilica on a Polyethylene System. *Journal of Physics: Conf Ser*, 1–6 (2011).

15. Martinez, J. M.; Augullo, T. G.; Juan, C. M.; Fernandez, G. A.; Barcelo, C. O.; Palau, A. T. Properties of Solvent based Polyurethane Adhesives Containing Fumed Silica. *Macromolecular Symposia.* 108(1), 269–272 (1996).

16. Mould, S.; Barbas, J.; Machado, A.V.; Nóbrega, J. M.; Covas, J. A. Measuring the Rheological Properties of Polymer Melts with on-line Rotational Rheometry. *Polym. test.* 30, 602–610 (2011).

17. Panaitescu, D. M.; Frone, A. N.; Spataru I. C. Effect of Nanosilica on the Morphology of Polyethylene Investigated by AFM. *Compos. Sci. Technol.* 74,131–138 (2013).

18. Preghenella, M.; Pegoretti, A.; Migliaresi, C. Thermo Mechanical Characterization of Fumed Silica-Epoxy Nanocomposites. *Polymer.* 46, 12065–12072 (2005).

19. Robinson, A. M.; Haworth, B.; Birley, A.W.; Elongational Rheometry of Polyethylene Terephthalate/Bisphenol-A Polycarbonate Blends. *Eur. Polym. J.* 32(9), 1061–1066 (1996).

20. Rosales, J. A.; Arias, G.; Rodrõ-Âguez, O. S.; Norman, S. A. Viscosity Changes Associated with the Chemically Induced Cross Linking of Plasticized Poly(vinyl chloride) Measured by Parallel Plate and Torque Rheometry: Influence of Magnesium and Barium Mercaptides. *Polym. Degrad. Stabil.* 68, 253–259 (2000).

21. Seong, H. K.; Seon, H. A.; Toshihiro, H. Crystallization Kinetics and Nucleation Activity of Silica Nano Particle filled Poly(ethylene 2,6-naphthalate). *Polymer* 44, 5625–5634 (2003).

22. Yang, F.; Ou, Y.; Yu, Z. Polyamide 6/Silica Nanocomposites Prepared by in situ Polymerization. *J. Appl. Polym. Sci.* 69 (2), 355–361 (1998).

CHAPTER 5

RATIONAL DESIGN OF MOLECULARLY IMPRINTED POLYMERS: A DENSITY FUNCTIONAL THEORY APPROACH

SHASHWATI WANKAR, AAKANKSHA JHA, and
REDDITHOTA J. KRUPADAM

*National Environmental Engineering Research Institute, Jawaharlal
Nehru Marg, Nagpur 440020, India; Tel.: +91-712-2249884;
Fax: +91-712-2249896; E-Mail: rj_krupadam@neeri.res.in*

CONTENTS

Abstract 77
5.1   Introduction 78
5.2   Materials and Methods 81
5.3   Results and Discussion 87
5.4   Conclusions 95
Acknowledgments 95
Keywords 96
References 96

ABSTRACT

The molecular imprinting technology that has recently demonstrated great potential for producing artificial receptors that challenge their natural counterparts. The stability and low cost of molecularly imprinted polymers (MIPs) make them advantageous for application as sensory materials, immunosorbents and adsorbents in environmental and biomedical fields. However, the imprinted polymer properties such as selectivity, capacity and binding kinetics towards the target molecule primarily depends on polymer composition and conditions followed during molecular imprinting. Availability of huge number of functional and cross-linking monomers, it would be time consuming as well as intense quantities of materials/reagents are required to select more appropriate polymer composition based on experiments for a given molecule. To overcome this constraint, the rational design using computer simulations has recently emerged as an efficient and experimental free way of selection of suitable polymer precursors to achieve the optimum molecular recognition properties of imprinted polymers. In this article, a new combinatorial screening method was proposed based on density functional theory (DFT) for selection of polymer precursors for microcystin-LR specific. The study also discusses about on the nature of intermolecular interactions responsible for high selectivity for the microcystin-LR (Scheme 1).

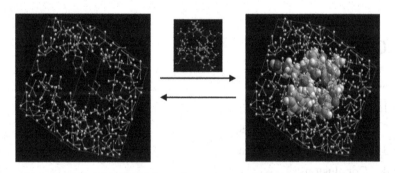

Microcystin-LR Imprinted Polymer   *Microcystin-LR specifically adsorbed in imprinted polymer*

SCHEME 1   Computationally designed microcystin-LR imprinted polymer and its view of adsorption–desorption phenomena.

5.1   INTRODUCTION

One of Nature's most important talents is evolutionary development of systems capable of performing various tasks, such as molecular recognition.

Over the past four decades, researchers have been inspired by nature to produce range of functional materials by design rather than by evolution. Now, the time is ripe to capture momentum on fundamental achievements of materials development and use developed materials and technology in real life applications serving growing industrial needs and public services. Some of the environmental research thrust areas are (i) development of nanometer sized materials and nanopatterned systems as super-sorbent polymers for separation of anionic and cationic organic compounds from environmental systems, (ii) photocatalytic materials for rapid degradation of recalcitrant molecules, (iii) functionalized materials for carbon dioxide sequestration, (iv) catalytic systems for reduction on nitrate in drinking waters and (v) biomimetic materials for odorous compounds detection and treatment. In past decades, polymeric adsorbents, in particular, emerged as highly effective alternatives to traditionally used materials such as activated carbon for pollutants removal from industrial effluents and drinking water (Savage and Diallo, 2005; Ali and Gupta, 2007; Ali, 2012). More recently, the development of polymer based hydrid adsorbents has opened up the new opportunities of their application in remediation and sensing of trace contaminants and heavy metals. A precisely designed material capable of selective recognition of targeted pollutants, in a fashion of antigen-antibody recognition in biological systems, is of prime importance in trace contaminants detection and remediation. One technique that is being increasingly used to prepare such highly selective polymer materials is "Molecular Imprinting."

Molecular imprinting involves the preparation of a solid material (usually a synthetic polymer) containing cavities that have a shape and functional groups complimentary to the imprinted template molecule (Wulff, 1995; Mosbach and Romstrom, 1996; Mathew-krotz and Shea, 1996). In general, MIPs are prepared through a simple process. First, polymer precursors are chosen based on the application of interest. Functional monomers that exhibit chemical functionality designed to interact with the template molecule via covalent or non-covalent interactions are selected. Also, the type and amount of cross-linking monomer that will provide structural support to the polymer network as well as define the pore size of diffusion of imprint molecules in to and out of the polymer matrix are selected. Once chosen, these precursors are dissolved, with the template, in an appropriate solvent.

Second, a pre-polymerization complex is formed between the imprint molecule and functional monomer, which forms the basis of the specific binding sites. The monomer and template mixture is then polymerized typically via free-radical polymerization and, finally, the imprint molecule is removed, which leaves a polymer network with three-dimensional binding cavities based on the imprint molecule of interest (Fig. 5.1). The molecularly imprinted polymers (MIPs) offer the positive advantages of low cost and high stability due to their chemical origin. Moreover, their synthetic nature readily allows the incorporation of integrated signaling functionalities, modification of surface chemistry and attachment of various labels during synthesis without affecting molecular recognition. Progress in computational methods has also made the design of MIPs and the selection of functional monomers a routine and reliable process (Batra and Shea, 2003; Khan et al., 2012). In this article, a new theoretical approach based on density functional theory (DFT) is reported to select appropriate polymers precursors for preparation of a model template, microcystin-LR. Microcystin-LR is the most toxic and primarily detected congener belongs to frequently occurring class of cyanobacterial toxins. Due to adverse health effects, the World Health Organization (WHO) established a provisional concentration limit of 1 ug L^{-1} for MC-LR in drinking water (WHO, 2003) and the United States Environmental Protection Agency (USEPA) has placed MCs on the Drinking Water Contaminants List. Reduction of threats to human health and aquatic life involves toxic cyanobacteria blooms and/or MCs to be monitored and removed from water columns, in particular, public water supplies.

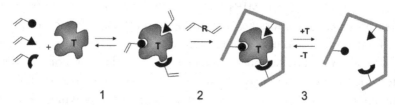

FIGURE 5.1 A schematic representation of molecular imprinting. After polymerization (step 1), the removal of template (step 2) creates a specific shape and size binding site in the polymer (Sellergren et al., 2005).

5.2   MATERIALS AND METHODS

5.2.1   COMPUTER SIMULATIONS

5.2.1.1   Workstation Used for Computer Simulations

The workstation Dell Precision T7500 used to simulate functional mono-mer–template interactions. The workstation was equipped with Intel (R) Xeon (R) CPU, X5660@2.80 GHz, 2.79 GHz Duel Processor running a 64-bit operating system and 24 GB RAM. This workstation was used to run the software Gaussian 4.1 ver.

5.2.1.2   Molecular Modeling and Simulations

The functional monomers in the library have a wide range of functional-ities, capable of forming both ionic and hydrogen bonds.

Initially, functional monomers and template, microcystin-LR used in the study were built and configures (geometrically optimized) and then using DFT approach in Becke 3-Parameter Exchange-Correlation Functional (B3LYP) level with 6–31G* basis set binding energies between each functional monomer and microcystin-LR were computed. A scheme showing the molecular modeling is depicted in Fig. 5.2.

The force fields and spatial configurations of functional monomers and microcystin-LR in the solvent (dichloromethane) were derived based on the parameter energy conformation. All the species were parameterized using procedure described by Seminario (2000) on the basis of quantum chemistry calculation (Gaussian 03, B3LYP/6–31G* set), where bonding force parameters were estimated from analysis of Hessian force matrix and atomic electric charges from ESP fit.

In the first step, MD simulations were applied to screen the best mono-mers targeted against microcystin-LR in acetonitrile; then, DFT simula-tions were conducted to calculate the interaction energies between the combinatorially screened functional monomers and microcystin-LR. Molecular Dynamics simulations were carried out using Gaussian 03 package and Optimized Potentials for Liquid Simulations (OPLS) force field (Cramer and Truhlar, 1999).

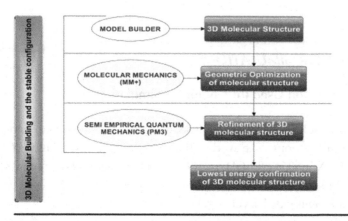

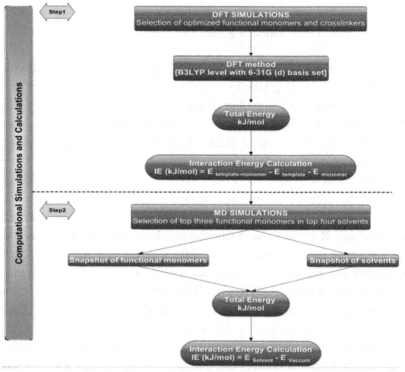

FIGURE 5.2 Computational Approach – An algorithm for computer simulations of microcystin-LR imprinted polymer.

The criterion for choosing the functional monomers to the template is the binding energy, which was calculated by the formula as follows (Eq. (1).

$$\Delta E_{MD/DFT} = E_{solvent}(template - monomer)$$

$$- E_{solvent}(template) - E_{solvent}(monomer) \qquad (1)$$

where, E (template–monomer) is the potential energy of the simulated system (monomer, template in solvent), E solvent (template) is the potential energy of template in solvent, and E (monomer) is the potential energy of monomer in solvent.

The virtual solvent box was prepared using a specific number of solvent molecules based on the expression (Eq.2):

$$\text{Number of molecules} = \frac{(\text{density} \times \text{volume} \times \text{Avogadro's number})}{\text{molecular weight}} \qquad (2)$$

Once the solvent box was constructed, the functional monomer and template molecules were placed in the periodic boundary conditions to determine the interaction energies (ΔE) through MD simulation. The box 18 × 18 × 18.6 nm (periodic boundary conditions in all three dimensions) contained 664 dichloromethane molecules, one microcystin-LR molecule and 6 molecule of functional monomer (IA). The MD simulation of molecular systems was performed in the NVT (number of molecules N, volume V and temperature T) ensemble was kept constant.

The bond lengths and the linear structure of the function monomers and template microcystin-LR were kept by using linear constraint solver (LINCS) algorithm. The algorithm is inherently stable, as the constraints are self-reset instead of derivatives of the constraints, thereby eliminating drift. Although the derivation of the algorithm is presented in terms of matrices, no matrix multiplications are needed and only the non-zero matrix elements are stored, making the method useful for very large molecules. During pre-polymer complex formation, it was hypothesized that each site interacts with all sites of different molecules with Lennard-Jones and Coulomb interactions. The cutoff distances of columbic and van-der-Waals force used to describe the interaction were 0.9 nm and 1.0 nm respectively. The employed Lennard-Jones parameters are σ = 3.40 Å and ε = 0.086 kcal/mol for carbon and σ = 2.60 Å and ε = 0.015 kcal/mol for hydrogen, which corresponds to values from the standard GAFF force field. Simulations were maintained with a time step of 0.5 fs for 2000 ps trajectories.

The Particle Mesh Ewald (PME) summation method was used to calculate the electronic interaction among the molecules in the system. On the other hand the role of solvent in polymerization was studied using Polarized Continuum Model (PCM) in which a numerical representation of the polarization of the solvent provides best computation of the electronic energy and solvent energy (Toukimaji et al., 1999; Khan and Krupadam, 2013). In order to better investigate the hydrogen bond formation and binding strength of hydrogen bonded complexes, Mullikan atomic charge distribution was determined. These computations provided information about spatial arrangement of functional monomers and the distances between the atoms of reactivity in the pre-polymer complex. Based on this information, the energy scores of polymer precursors are useful in selection of appropriate recipe for preparation of MIP for microcystin-LR. The analysis tools for studying the energy and trajectory of the simulated system were facilitated with the Gaussian 4.1 software.

5.2.2 PREPARATION OF MIPS

5.2.2.1 Chemical and Reagents

The template, microcystin-LR was purchased from Alexis Biochemicals (San Diego, USA). The functional monomers itaconic acid (IA), methacrylic acid (MAA), and 2-Acrylamido-2-methyl-1-propanesulfonic acid (AMPSA) and cross-linking monomer ethylene glycol dimethacrylate (EGDMA) were purchased from Sigma-Aldrich (Bachs, Switzerland); while the solvents dichloromethane, chloroform, acetonitrile, dimethyl sulfoxide, methanol and trifluoroacetic acid were procured from Merck (Darmstadt, Germany). 2,2'-azobisisobutyronitrile (AIBN) was brought from Acros Organics (Geel, Belgium). The molecular structures of chemicals used in the study are shown in Fig. 5.3. The monomers were purified prior to use via standard procedures in order to remove stabilizers. The AIBN was recrystallized from acetone and the acetonitrile

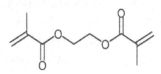

Microcystin-LR

Functional Monomers

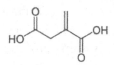

Methacrylic acid

Ethylene glycol dimethacrylate

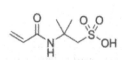

Itaconic Acid

Azobisisobutyronitrile

Dichloromethane

2-Acrylamido-2-methyl-1-
propanesulfonic acid
(AMPSA)

FIGURE 5.3 Molecular structures of polymer precursors used in the preparation of microcystin-LR imprinted polymers.

dried over molecular sieves. All reagents were either HPLC grade or analytical grade. Deionized water was obtained by passing tap water through a Milli-Q system with the conductivity <15 MΩ cm-1 and on-line TOC <5 μg L-1. The stock solution of microcystin-LR was prepared at the concentration of 10 μg mL-1 in deionized water containing 0.1% trifluoroacetic acid.

5.2.2.2 Preparation of MIPs

The template, microcystin-LR (9.93 mg, 1.0 μmol) was dissolved in 5 mL dichloromethane in a 10 mL glass vial. To this, the functional monomer itaconic acid (630 mg, 6.0 μmol) was added and then the glass vial was placed in a refrigerator at 0°C for 30 min. Later, 1.5 mmol (991 mg) cross-linking monomer EGDMA and 10 mg AIBN were added to the solution. The sealed glass vial containing reaction mixture was freeze- thaw-degassed by submerging the tube in liquid nitrogen and holding the frozen tube under a vacuum of 100 mTorr for a period of 15 min. The tube was then sonicated for 5 min and placed in water bath at 40°C for 16 h. Upon completion of polymerization, the tube was taken out of the water bath and crushed. The polymer monolith was ground in a ball mill to polymer particles of size 75 μm or less (200 mesh). The template MC-LR was extracted in batch mode, using methanol on a horizontal shaker for 24 h. The washing procedure was repeated (10 times) until MC-LR in the extraction solvent could not be detected by LC/MS. Finally, the particles were dried under vacuum

TABLE 5.1 Composition of Microcystin-LR Imprinted and Non-Imprinted Polymers

	Template, Microcystin-LR (μmol)	Functional monomers (μmol)			Cross-linking monomer (mmol)	Solvent (DCM), mL
		IA	MAA	APMSA	EGDMA	
MIP-1	1.0	6	-	-	1.5	5
NIP-1	-	6	-	-	1.5	5
MIP2	1.0	-	6	-	1.5	5
NIP-2	-	-	6	-	1.5	5
MIP-3	1.0	-	-	6	1.5	5
NIP-3	-	-	-	6	1.5	5

MIP, molecularly imprinted polymer; NIP, non-imprinted polymer; IA, itaconic acid; MAA, methacrylic acid; AMPSA, 2-Acrylamido-2-methyl-1-propanesulfonic acid; DCM, dichloromethane for further use. Apart from this MIP, two different MIPs were prepared with using methacrylic acid (MAA) and 2-Acrylamido-2-methyl-1-propanesulfonic acid (AMPSA) functional monomers. The composition of polymers was given in Table 5.1. The corresponding non- imprinted polymers (NIPs) was prepared in parallel in the absence of MC-LR and treated in the same manner.

5.3   RESULTS AND DISCUSSION

5.3.1   COMBINATORIAL SCREENING OF FUNCTIONAL MONOMERS

The most suitable functional monomer for the targeted analyte microcystin-LR was chosen based on the interaction energy between microcystin-LR and the functional monomers computed on MM+, PM3 and DFT theories. A virtual library of microcystin-LR and functional monomer complexes was prepared based on interaction energy scores (Table 5.2). The library data showed that the functional monomers – itaconic acid (IA), methacrylic acid (MAA), and 2-Acrylamido-2-methyl-1-propanesulfonic acid (AMPSA) form better interactions with microcystin-LR; the highest interaction energy (ΔE) of microcystin-LR with functional monomers favors to form the most stable complexes in the equilibrium state, and such combination would be more suitable for preparation of MIPs. The computational data concludes that the best order of functional monomers ΔE (IA) > ΔE (AMPSA) > ΔE (MMA), indicating that microcystin-LR interacts strongly with IA and is the most suitable functional monomer for microcystin-LR imprinted polymer preparation from the list of 24 functional monomers.

The optimum ratio of functional monomer and microcystin-LR was determined using the interaction energies scores computed by adding functional monomer molecules to the microcystin-LR where the leap in binding energy of the complexes of microcystin-LR and functional monomer was considered. The microcystin-LR with each of the monomers at a molar ratio of 1:1, 1:2, 1:3, ..., 1:N was computed by applying the conformation optimization to these complexes. From the Table 5.3, the theoretical titration results showed that the six IA molecules give the highest interaction energy score, that is, −1997.62 kcal/mol and 1:6 is a

TABLE 5.2   Binding Energy Between MC-LR and Functional Monomers Computed in DCM Solvent System on Gaussian 03W Software

Template	Functional Monomers	Binding energy, Kcal mol^{-1}
MC-LR	1-vinylimidazole	−470.68
	2 (5)-Vinylimidazole	−217.84
	2-vinylpyridine	−479.59
	4-ethylstyrene	−488.01
	4-vinylpyridine	−460.75
	Acrylamide	−123.68
	Acrylamido-2-methyl-1-propane- sulphonic acid	−521.17
	Acrylic acid	−429.28
	Acrylonitrile	−214.93
	Allylamine	−316.71
	Itaconic acid	−611.78
	Methacrylamide	−485.90
	Methacrylic acid (MAA)	−512.85
	Methyl Methacrylic acid (MMA)	−366.36
	N-(2-aminethyl)-methacrylamide	−429.73
	N, N, N,-trimethyl aminoethyl methacrylate	−218.70
	N, N'-diethyl-4-styrylamidine	−389.04
	N, N'-diethyl aminoethyl methacrylamide (DEAEM),	−425.71
	N-vinylpyrrolidone (NVP)	−354.94
	p-vinylbenzoic acid	−437.08
	Styrene	−376.43
	Trans-3-(3-pyridyl)-acrylic acid	−411.45
	Trifluoro methacrylic acid (TFMA)	−450.73
	Urocanic ethyl ester (UEE)	−490.26

more appropriate template and functional monomer ratio to form effective binding sites in the MIP. In this work, four solvents were selected based on the criteria – polarity. Considering the solvent effect on prepolymer complex formation, MD simulations were performed in explicit solvent to achieve the more accurate physical description of complex

TABLE 5.3 Binding Energy Computed by the Addition of IA Molecules to MC-LR in a Virtual Solvent System

Number of IA molecules	1	2	3	4	5	6	7	8
ΔE (binding energy), Kcal/mol	−611.78	−707.74	−806.92	−994.12	−1264.88	−1997.62	−1440.51	−938.64

ΔE, binding energy between IA and microcystin-LR was computed on DFT simulations of B3LYP Level with 6031G(d) basis set. The virtual solvent system has dimension of 18x18x18 nm containing 664 molecules of solvent, DCM. Each binding energy value is the average of 10,000 iterations on the software Gaussian.

system, producing a trajectory of the molecules at the time scale of 1 ns on Gaussian 03 software. From the snapshot, it was found that the polar protic solvents give the positive interaction energy due to existence of hydrogen bonding. This leads to the possibility of forming the more hydrogen bonding with the functional monomer during the imprinting. The binding distances between microcystin-LR and functional monomers were computed for different solvent systems in addition to interaction energies. The bond distances data provides low strong/weak interaction exists between microcystin-LR and functional monomers (Fig. 5.4). It was found that among the solvents, the increasing order of interaction energy simulated follows: dichloromethane (−1997.62 kcal/mol) > chloroform (−1675.51 kcal/mol) > acetonitrile (−1527.50 kcal/mol) > dimethylsulfoxide (−1469.54 kcal/mol). The results represents that dichloromethane is most favorable polymerization solvent for the MIP using polymer recipe, that is, microcystin-LR and IA.

5.3.1.1   Investigations on Pre-Polymer Complex (PPC) Stability Using DFT Computations

The stability of the pre-polymer complex formed between microcystin-LR and the functional monomer during imprinting mainly influenced by

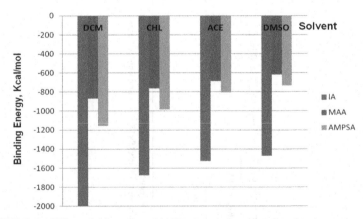

FIGURE 5.4   Effect of solvents on binding energy between functional monomers and microcystin-LR (DCM, dichloromethane; CHL, chloroform; ACE, acetonitrile; DMSO, dimethyl sulfoxide).

bonding mechanism. Hydrogen bonding interaction plays a significant role in the formation of pre-polymerization complex (PPC). In most of the cases, the functional group of monomers interacting with template are either –COOH or CH =CH–. In the present case-COOH group of IA form the closest binding distance 1.97 Å in DCM. The binding distance between microcystin-LR and IA was computed for other solvents (chloroform) CHL, acetonitrile (ACN) and dimethylsulfoxide (DSMO) were 2.17, 3.26, and 3.89 Å, respectively. It is generally considered that the hydrogen bonding occurs as donor-acceptor distance is smaller than 2.5 Å. Hence, IA has H-bonding interaction with microcystin-LR in dichloromethane. The binding energy computed for microcystin-LR and IA was –661 kcal/mol) and the binding distance (1.97 Å) represents that the dominant bonding would be H-bonding; however, the functional group of IA also form π-π-bonding in DCM. For better understanding of interaction between IA and microcystin-LR at electronic level, the electronic structure of the microcystin-LR were treated with the B3LYP/6–31G (d, p) level to derive the Mulliken charges distribution to pin-point sites of interaction. From the partial charges on the molecules and the spatial considerations, it was derived that the proton donor is hydrogen atom of carboxylic group in IA possessing with the charge of +0.107e in acetonitrile (the hydrogen atoms in carboxylic group of IA has just one value of partial charge due to molecular symmetry), while that of 1:1 microcystin-LR/IA pair is +0.112e. The positive Mulliken charge of hydrogen atom increases due to the de-shielding effect of protonation by the oxygen atom in IA whose charge decreases from -0.284e to -0.306e indicating the down-field shift of the carboxylic proton. Usually the angle θ between the donor H and H⋯acceptor bonds is in the range from 140° to 180°, where smaller angle values represent weak hydrogen bond. The θ between O⋯H bond in carboxylic group of IA and H⋯O bond is 157° demonstrating week hydrogen bond interactions exists between IA and microcystin-LR in acetonitrile solvent during polymerization.

It can be seen that the oxygen and nitrogen atoms may participate in π-π interactions and H-bonding with the proton on the hydroxyl group. The formation of the complex between microcystin-LR and functional monomer IA is the key step in the preparation of a MIP selective for the cyanotoxin. The interaction energy (E) between microcystin-LR and IA

(−41.69 kJ/mol) was calculated in dichloromethane solvent. The theoretical results shows that the solvent- dichloromethane has the high affinity to both the template molecule microcystin-LR, and the monomer IA and cross-linker EGDMA. The charge contours of cyanotoxins with functional monomer IA were depicted in Fig. 5.5a, b. This affinity acts to shield or reduce the interaction between microcystin-LR and functional monomer, which is required in the formation of the microcystin-LR – IA of the complementary structure prior to the initiation of polymerization.

5.3.1.2 Comparison of Theoretical and Experimental Studies

The theoretical computations were given information on most suitable polymer precursors for preparation of microcystin-LR imprinted polymer. The computations also provided suitable the molar ratio between microcystin-LR and functional monomer and also information about the nature of bonding between the microcystin-LR and functional monomers (IA, MAA and APMSA). Complimentary to the theoretical predictions, three sets of molecular imprinted polymers (IA-MIP, MAA–MIP and APMSA–MIP) were prepared from the same composition predicted from the theoretical computations. The adsorption capacity of three

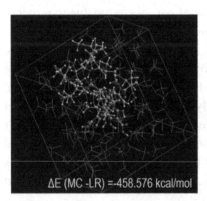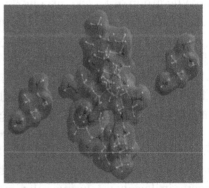

ΔE (MC -LR) =-458.576 kcal/mol

FIGURE 5.5 (a) 3D visualization of microcystin-LR with functional monomer itaconic acid in dichloromethane solvent (b) Charge density contours of microcystin-LR and itaconic acid in simulated solvent box with vacuum (the affinity evaluated showed that the interaction between microcystin-LR and functional monomer (IA) is the complementary structure prior to the initiation of polymerization, which is an important step in molecular imprinting).

TABLE 5.4 Binding Capacity of the Imprinted and Non-Imprinted Polymers

Imprinted/non-imprinted polymer	MIP-1	NIP-1	MIP-2	NIP-2	MIP-3	NIP-3
Binding Capacity, mg/g	21.67±1.46	3.62±0.17	14.35±1.33	3.59±0.11	17.54±1.41	2.75±0.12

Binding capacity experiments of MIPs/NIPs were performed in batch mode. The initial concentration of microcystin-LR was 10 mg/L; and 50 mg of each MIPs or NIPs was used in 10 mL of microcystin-LR aqueous solution. Contact time maintained was 30 min. The experiments were repeated for 5 times (n=5) under identical experimental conditions.

types of MIPs was determined and found that the binding capacity of the IA–MIP was higher than the MAA–MIP and the APMSA–MIP. The experimental results of adsorption capacities of three MIPs are given in Table 5.4. The effect of solvents on the microcystin-LR binding capacity was evaluated using four solvents (acetonitrile, chloroform, dichloromethane and dimethylsulfoxide). The results represents that the high adsorption capacity was achieved with the MIP prepared in dichloromethane and the lowest adsorption capacity for MIP synthesized in acetonitrile. These experiments verify that, according to the theoretical predictions, among different monomers and solvents tested, IA is the best functional monomer and dichloromethane is the most favorable solvent to prepare an imprinting polymer to selectively bind microcystin-LR. The relationship between theoretically derived ΔE and experimentally determined binding capacity was established (Fig. 5.6); the relationship is a linear with r=0.931. These few experiments confirm the application of computer simulations for combinatorial screening of

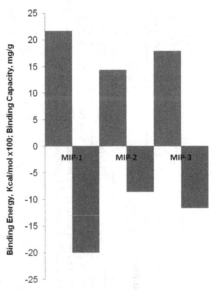

FIGURE 5.6 Comparison of theoretical study (in terms of binding energy) and experimental investigations (binding capacity of MIPs) (the data showed that there is a linearity between theoretical and experimental studies with a correlation value of r^2=0.931).

polymer precursors for preparation of effective molecular imprints for the targeted molecules.

5.4 CONCLUSIONS

The computational approach used in this article able to prepare highly selective polymer for microcystin-LR from aqueous solutions. The following important conclusions were derived from the study:

(i) Molecular dynamics simulations were carried out to screen best functional monomers from the library of 24 functional monomers; and the functional monomer itaconic acid (IA) showed the best binding energy score (-188.1211 Kcal mol^{-1}) for the targeted molecule microcystin-LR.

(ii) The binding interactions between functional monomer and the targeted molecule (microcystin-LR) was computed the DFT method shown the existence of H-bond and π-π interactions in the transient pre-polymer complex formation. The Mullikan charge analysis data also supported the nature of binding between the functional monomer and the microcystin-LR.

(ii) Experimentally, the three MIPs were prepared with different polymer precursors formulations and IA-EGDMA formulated MIP showed highest binding capacity for microcystin-LR, which is acceptable as per the computer simulations.

(iv) The results reported in this article provide an insight into better understanding of pre- polymer complex formation during molecular imprinting at molecular level; and help in design of molecular imprints for different applications.

ACKNOWLEDGMENTS

The Council of Scientific and Industrial Research (CSIR) and Planning Commission, Government of India supported this work under the project "Molecular Environmental Science and Engineering Research" (MESER).

KEYWORDS

- Becke 3-Parameter Exchange-Correlation Functional (B3LYP) level
- density functional theory
- linear constraint solver (LINCS) algorithm
- methacrylic acid
- molecularly imprinted polymers
- optimized potentials for liquid simulations
- Particle Mesh Ewald (PME) summation method
- Polarized Continuum Model
- World Health Organization

REFERENCES

1. Ali, I. New Generation Adsorbents for Water Treatment. *Chem. Rev.* 112, 5013–5091(2012).
2. Ali, I.; Gupta, V.K. Advances in Water Treatment by Adsorption Technology. *Nat. Protocols.* 1, 2661–2667(2007).
3. Batra, D.; Shea, K.J. Combinatorial Methods in Molecular Imprinting. *Curr. Opin. Chem Biol.* 7, 434–442 (2003).
4. Cramer, C.J.; Truhlar D.G. Implicit Solvation Models: Equilibrium, Structure, Spectra and Dynamics. *Chem. Rev.* 99, 2161–2200 (1999).
5. Hall, A.J.; Emgenbroich, M.; Sellergren, B. (2005); Hall, A. J.; Emgenbroich, M.; Sellergren, B. (2005): Imprinted Polymers, in Templates in Chemistry II, *Topics in Current Chemistry, C.A. Schalley, K.-H. Dötz, F. Vögtle (Eds.), Springer-Verlag, Heidelberg, Germany*, 249, 317–349.
6. Khan, M.S.; Krupadam R.J. Density Field Theory Approach to Design Multi-template Imprinted Polymer for Carcinogenic PAHs Sensing. *Comb. Chem. High Throughput Screen.* 16, 682–694 (2013).
7. Khan, M.S.; Wate, P.S.; Krupadam R.J. Combinatorial Screening of Polymer Precursors for Preparation of Benzo[a]pyrene: an *ab initio* Computational Approach. *J. Mol. Model.* 18, 1969–1981 (2012).

8. Mathew-Krotz, J.; Shea, K.J. Imprinted Polymer Membranes for the Selective Transport of Targeted Neutral Molecules. *J. Am. Chem. Soc.* 118, 8754–8755 (1996).

9. Mosbach, K.; Ramstrom, O. The Emerging Technique of Molecular Imprinting and Its Future Impact on Biotechnology. *Nat. Biotechnol.* 14, 163–170 (1996).

10. Savage, N.; Diallo, M.S. Nanomaterials and Water Purification: Opportunities and Challenges. *J. Nanoparticle Res.* 7, 331–342 (2005).

11. Seminario, J.M.; Zacarias, A.G.; Tour, J.M. Theoretical Study of a Molecular Resonant Tunneling Diode. *J. Am. Chem. Soc.* 122, 3015–3020 (2000).

12. Toukmaji, A.; Sagui C.; Board, J.; Dardeu T. Efficient Particle Mesh Ewald based Approach to Fixed and Induced Dipolar Interactions. *J. Chem. Phy.* 113, 10713–10725.

13. WHO (2003) Cyanobacterial toxins: Microcystin-LR in drinking-water. Background document for preparation of WHO Guidelines for drinking-water quality. Geneva, World Health Organization (WHO/SDE/WSH/03.04/57).

14. Wulff, G. Molecular Imprinting in Cross-linked Materials with the aid of Molecular Templates – A way towards Artificial Antibodies. *Angew. Chem. Ing. Ed. Engl.* 34, 1812–1832 (1995).

CHAPTER 6

PREPARATION OF POLYMER AND FERRITE NANOCOMPOSITES FOR EMI APPLICATIONS

P. RAJU and S. R. MURTHY

Department of Physics, Osmania University, Hyderabad–500 007, India

CONTENTS

Abstract 99
6.1 Introduction 100
6.2 Synthesis of Polymer/Ferrite Nanocomposites 103
6.3 Synthesis of Pani + Nicuzn Ferrite Nanocomposites 105
6.4 Experimental Results and Discussions 108
6.5 Conclusions 117
Keywords 118
References 118

ABSTRACT

Polymer-ferrite nanocomposites for electromagnetic interference (EMI) shielding are reviewed. The polyaniline-on-$Ni_{0.48}Cu_{0.12}Zn_{0.4}Fe_2O_4$ nanocomposites is taken as example and discussed. The Polymer-NiCuZn ferrite with high dielectric absorbing properties and electromagnetic shielding effectiveness at low frequencies were successfully prepared through an in situ emulsion polymerization method. Polyaniline (PANI) was added with hydrochloric acid to improve its electrical properties and interactions with ferrite nanoparticles. The ferrite nanoparticles are assembled on the polyaniline surface and improved the thermal stability of the polyaniline. Polyaniline nanofibers were found to have an average diameter of 60 nm and length of 100 nm. Similarly, the ferrite nanoparticles were with spherical shape with an average diameter 30 nm. The complex permittivity and shielding effectiveness of the ferrite-grafted polyaniline nanocomposites increased with the increasing weight percentage of polymer. From the above studies it was found that all these nanocomposites were suitable for the fabrication EMI devices.

6.1   INTRODUCTION

6.1.1   NANOCOMPOSITES BASED ON POLYMER-FERRITE

Nanocomposites of polymers are potentially important and have number of significant advantages over traditional polymer composites. The micron-sized composites usually require a high content of the filler phase to achieve the desired properties of the composite material. But, the nanocomposites can achieve the desired and better properties with a much smaller amount of the filler. This will yield materials with lower density and higher processability (Bhat and Geetha, 1999; Bins and Associates, 2002). The nanoparticles have advantages over micron-sized particles, for example, because they can exhibit novel magnetic, optical, thermal, electrical and mechanical properties. Similarly, in nanoparticles, the magnetic properties, such as, coercivity, saturation magnetization, and frequency dependent permeability are very different from those found in micron-sized materials. The polymers provide a processable matrix in which to disperse the particles.

The composites of polymer have many applications in many fields such as, automobile industry, the medical field as various types of sensors. General Motors is the first industry used a polypropylene/clay nanocomposite to fabricate the step-assist for some of its vans. The part was previously made from conventional talc-reinforced material. By replacing this part of the conventional micron-sized composite with that of nanocomposite materials resulted in increased stiffness, ductility at low temperatures and enhanced appearance. A weight savings of more than 20% was also achieved for material being replaced by the polymer nanocomposite. This was the first commercial automotive application for a polymer nanocomposite material based on a polymer such as polypropylene, polyethylene, or polystyrene. Many automotive companies are currently using polymer nanocomposites in the fuel lines and fuel system components.

In general, various types of metals, ceramics, and polymers are considered suitable for use in biomedical applications (Nam et al., 2009). However, they possess many disadvantages and these make alternative materials are desirable. Metals have many disadvantages such as, corrosion, high density, much higher stiffness compared to tissues, release of metal ions, which may cause allergic reactions, and low biocompatibility. Ceramics possess many problems, which include brittleness, low fracture strength, lack of resilience, and low mechanical reliability. Comparative to the metals and ceramics polymers are too flexible and too weak to be used in certain applications. Apart from this their properties can be adversely affected by sterilization processes and they also absorb liquids and swell, or leach undesirable products. Whereas polymer nanocomposites are currently used in various medical applications and many additional applications have been proposed. For example, repair or replacement of bones, bone plates and screws cartilage, tendons, ligaments, muscles, finger joints, hips, knees, abdominal walls, vascular grafts, dental composites, and spine cages, plates, rods, screws, and discs. Nanocomposites of polymers are excellent candidates for different types of sensing applications. This is due the fact that the conductive and absorptive properties of insulating polymers doped with condwucting materials, and the absorptive properties of insulating polymers with non-conducting fillers are sensitive to exposure to gas vapors. Hence, they are useful to monitor the existence and concentration of gases in the environment. The electrical resistance of conductive polymer nanocomposites is changes when these materials are subjected to tensile strain and pressure. Because of this property of nanocomposites make them possible applications as pressure and stretch sensors for detection of deformations and vibrations of mechanical devices such as vehicle parts.

Polymer nanocomposites have many applications in the area of Electromagnetic interference (EMI) suppression. All electronic devices generate and emit radiofrequency waves that can interfere with the operation of electronic components within the same device as well as other electronic devices. In present day all electronic devices are moving towards miniaturization and in this process the electronic components are to be packed very close to each other, which increases the problem of electromagnetic interference. When an electromagnetic wave is incident upon a conductive surface, energy is reflected and absorbed. The ability of a material to shield electromagnetic energy, whether it be unwanted energy entering a system or escaping a system and it is called as it's shielding effectiveness (White, 1975). It is consists of losses due to absorption, reflection, and multifull reflections. EMI suppression over a wideband frequency range requires tunability of the impedance (Z), which depends on the tunability of the complex permeability and complex dielectric constant. The conductivity plays an important role in a material's ability to shield electromagnetic energy.

Metals, along with capacitors and ferrites are commonly used in EMI suppression and these materials have many disadvantages in terms of their weight, corrosion and physical rigidity. These properties could be overcome with the development of new composite materials. Out of all composite materials, conducting and insulating polymers doped with magnetic nanoparticles are lightweight, flexible alternatives to the micron sized metal components. Hence, polymer/ferrite nanocomposites are best choice. Another importance of polymer nanocomposites is that critical parameters such as loss tangents and impedance matching, which are important in microwave devices, may be controlled in these materials. The loss tangent is a measure of the inefficiency of a magnetic system. The loss tangents are primarily determined by magnetic and eddy current losses and which depend on the resistivity of the material. The resistivity of conducting polymer is increases with an addition of nanoparticles, thereby decreasing the eddy current losses. Generally, magnetic losses in composite materials are controlled by the material grain structure, domain wall resonances etc. and these parameters can be manipulated effectively in nanocomposite materials by the size distribution of nanoparticles and their dispersion in the host matrix.

During the transmission, impedance mismatch between source and load in circuits is the main source of signal attenuation. By manipulating the properties of nanocomposites desired impedance values at specific frequencies is attained and the attenuation can be minimized. An imped-ance of nanocomposite material can be adjusted by the type and amount of magnetic nanoparticles dispersed in the polymer. The combination of the ferrimagnetic nanoparticles with conducting polymer leads to forma-tion of an important nanocomposite possessing with unique combination of electrical and magnetic properties. This property of the nanocompos-ite materials can be used for utilizing them as electromagnetic shielding material. This is because, the electromagnetic wave consist of an electric (E) and the magnetic field (H) right angle to each other. The ratio between E to H factor (impedance) has been subjugated in the shielding applica-tion. In polymer/ferrite type of nanocomposites, the conducting polymer type of materials can effectively shield electromagnetic waves generated from an electric source, whereas electromagnetic waves from a magnetic source can be effectively shielded only by magnetic materials.

The primary mechanism of EMI shielding is usually reflection and for this type of the radiation by the shield, it must have mobile charge carriers (electrons or holes), which interact with the electromagnetic fields in the radiation. Hence, the shield tends to be electrically conducting, without a high conductivity. For example, a volume resistivity in the order of 1 Ωcm is typically used. A secondary mechanism of EMI shielding is usually absorption. For significant absorption of the radiation, the shield should have electric and/or magnetic dipoles, which interact with the electromag-netic fields in the radiation. Thus, having both conducting and magnetic components in a single system could be used as an EMI shielding material.

6.2 SYNTHESIS OF POLYMER/FERRITE NANOCOMPOSITES

Generally two methods are employed to prepare polymer nanocomposites and they are: physical and chemical methods. The physical methods of synthesis are solid blending (White, 1975), recrystallization from solution or suspension, polymer melt intercalation, spray coating, etc. In the chemi-cal methods mainly involve in situ polymerization. The main difference between chemical methods and physical methods is fabrication of poly-mer–inorganic nanocomposites by the mixing of the two phases before or after the polymer is formed. Generally, physical methods are easier to handle, chemical methods are capable of producing more stable and more homogeneous nanocomposites. This is due to inorganic nanofillers have stronger interaction with monomer than with already formed polymers, giving in a better dispersion.

Recently many studies on nanocomposites of polymer and magnetic nanoparticles have been reported. Out these studies, Burke et al. (2002) reported the preparation of polymer-coated iron core–shell nanoparticles and others reported on bulk PMMA matrix incorporated with magnetic nanoparticles. Fang et al. (2002) has reported the fabricated CoPt–PMMA nanocomposites. In this case, the magnetic nanoparticles were first synthesized and then dispersed into the monomer with the help of cross-linking agents and then followed by a polymerization process. But, Wilson et al. (2004) and Baker et al. (2004) reported the fabrication of Fe nanoparticles and Fe-oxide/Fe core/shell nanoparticles, respectively and dispersing them in a PMMA matrix by physical methods.

Recently, there has been a grownup interest in low temperature sintered NiCuZn ferrites for the application in producing Multilayer type Chip Inductors and microwave devices because of their better properties at high frequencies than Mn-Zn ferrites and the lower densification temperature than Ni-Zn ferrites. Apart from this, it exhibits different kinds of magnetic properties such as paramagnetic, superparamagnetic or ferrimagnetic behavior depending on the particle size and shape. Also, it exhibits unusual physical and chemical properties when its size is reduced to nano size. The absorbing characteristics of the materials depend on layer thickness, complex permittivity (ε^*), complex permeability (μ^*) and the frequency. It was found that the ferrite is such a kind of microwave absorbers with required complex permittivity and complex permeability (27,28). The core-shell structure composite nanoparticles exhibit improved physical and chemical properties and hence, they are very useful in a broader range of applications.

As mentioned above conducting polymer composites with micro/nanostructures have attracted significant academic and technological attention because of their potential applications in nanoelectronics, electromagnetics, and biomedical devices. Among these conducting polymers composites decorated with organic nanoparticles are of particular interest. This is because of possible interactions between the inorganic nanoparticles and the polymer matrices gives unique physical properties upon the formation of various micro/nanocomposites.

Out of all the conducting polymers, polyaniline (PANi) is the most versatile because of easier and inexpensive preparation methods. A part from this, they possess desirable properties, such as thermal and chemical stability, low specific mass, controllable conductivity and high conductivity at microwave frequencies. PANi is an important material with many potential applications in various fields such as electrical–magnetic shields, microwave absorbing materials, batteries, sensors and corrosion protections.

The fabrication of PANi/ferrite nanocomposite has been reported by using different methods such as in situ polymerization of aniline in the presence of $Zn_{0.6}$ $Cu_{0.4}$ $Cr_{0.5}$ $Fe_{1.4}O_4$ nanoparticles. The micro emulsion process was used for the preparation of PANi/NiZn ferrite nanocomposite. The oxidative electro-polymerization of aniline method was used in an aqueous solution in the presence of MnZn ferrite and NiMnZn ferrite. From the studies of electromagnetic properties on the PANi/ferrites, it was found that they were improved and tailored by controlling the addition of the ferrite in the composite. It was also found that the addition of ferrite to the PANi led to an increase in its thermal stability and decreased it electrical conductivity. Keeping these points in view, we have undertaken a detailed study of preparation and characterization of PANi-on-Ni CuZn Ferrite nanocomposites. The studies of microwave absorbing property on PANi-ferrite nanocomposite in the wide frequency range has been carried out and obtained results are presented in this chapter.

6.3   SYNTHESIS OF PANi + NiCuZn FERRITE NANOCOMPOSITES

The nanopowder of $Ni_{0.48}Cu_{0.12}Zn_{0.4}Fe_2O_4$ (NCZ) have been synthesized using microwave-hydrothermal method at a low temperature of 160°C/45min. starting with pure chemicals of (99.98%) Nickel nitrate [Ni $(NO_3)_2 \cdot 6H_2O$], (99.96%) Zinc nitrate [Zn $(NO_3)_2 \cdot 6H_2O$], (99.98%) Copper nitrate [Cu $(NO_3)_2 \cdot 6H_2O$] and (99.99%) Ferric nitrate [Fe $(NO_3)_3 \cdot 9H_2O$] are taken in stoichiometric ratio. These reagents were dissolved in 50 mL of de-ionized water. To this solution, sodium hydroxide (NaOH) was added to maintain the pH of the solution ~12. Controlling of pH is the key factor to synthesize the nano powder. Then the precipitation was transferred into double-walled digestion vessels that have an inner liner and cover made up of Teflon PFA and an outer high strength layer made up of Ultem polyetherimide and then treated using M-H method at 160°C/45 min. The M-H treatment was performed using a microwave accelerated reaction system (MARS-5, CEM Corp., Mathews, NC). This system uses 2.45 GHz microwave frequency and can be operated at 0–100% full power (1200±50 W). The reaction vessel was connected to an optical probe to monitor and control the temperature during synthesis. The product was separated by centrifugation and then washed repeatedly with de-ionized water, followed by drying in an Oven overnight at 60°C. Thus, the obtained powders were weighed and the percentage yields were calculated from the total expected based on the solution concentration and volume and the amount that was actually crystallized and a yield of 96% was obtained. Then the powders were characterized by XRD (PhilipsPW-1730 X-ray diffractometer), and TEM (model JEM-2010, JEOL, Tokyo, Japan).

Figure 6.1 shows the TEM picture for as synthesized NiCuZn ferrite (NCZ). The selected-area diffraction (SAD) patterns of these specimens reveal the formation of single-phase polycrystalline ferrites. No trace of an impurity or undesirable phase was detected in any of the ferrite. The diffraction rings, characteristic of nanocrystalline aggregates, have been indexed and obtained the average value of particle size is 30 nm. Occasional spots in the SAD pattern may arise from coarse crystallites or agglomerates. It may be noted that the observed average particle sizes from TEM picture are nearly same as that calculated from XRD peak broadening.

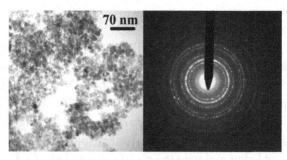

FIGURE 6.1   TEM picture for NiCuZn ferrite (NCZ).

Ferrite-on-polyaniline nanocomposites were synthesized by the in emulsion polymerization method. Figure 6.2 shows a experimental set up used for the synthesis. Solution of surfactant was prepared by dissolving 2.5 mL Tween-80 in 25 mL ethanol. Aniline (0.5 mL) was dissolved in 50 mL (1 M) HCl. The surfactant solution and aniline solution were mixed together under intense sonication to form a homogenous, milky colored emulsion. APS (0.58 g), which acted as an oxidant, was dissolved in 50 mL (1 M) HCl. Various amounts of ferrite nanoparticles (0.00 g, 0.10 g, 0.25 g, 0. 50g and 1.0 g) synthesized using microwave hydrothermal method, were added to the oxidant solution. Then, they were sonicated until they were dispersed uniformly in the solution. The dispersed in APS ferrite nanoparticles were added drop-wise in the aniline emulsion. Temperature of the reaction mixture was controlled below 280 K by using an ice bath. After the some time, the color of the mixture had transformed from milky white to dark green. Ferrite–polyaniline nanocomposite was obtained by centrifuging the green mixture at 20000 rpm/10 min, then washed with deionized water and acetone for five times, and dried at 330 K/2 h.

The prepared nanocomposites were characterized by X-ray diffraction (XRD) (Philips PW 1730), Transmission electron microscopy (TEM), Scanning Electron Microscopy (SEM) and Fourier Transform Infrared

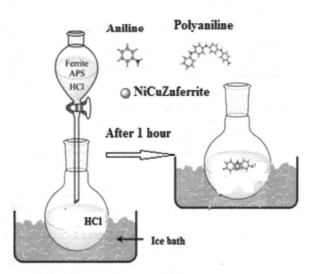

FIGURE 6.2   Experimental set up used for the preparation of PANi+ferrite nanocomposite.

spectrometer (FT-IR) (Brucker tensor 27). Crystallite size of the milled samples was calculated using Scherer equation, based on the widening of three strongest peaks of a certain phase. Particle size and morphology was investigated with a scanning electron microscope (SEM). The magnetic properties such as saturation magnetization (Ms) and coercive field (Hc) for the samples were obtained from the recorded hysteresis loops obtained with the help of VSM (model Lakeshore, Model 7300) at room temperature. The complex permittivity (ε^*), and permeability (μ^*) were measured in the frequency range of 1 MHz –1.8 GHz using LCR meter at room temperature. The shielding effectiveness (SE) is determined as a ratio of the power received without shielding material (P_0) to that of the power received with shielding material (P_1): SE = –10 log(P_0/P_1).

6.4   EXPERIMENTAL RESULTS AND DISCUSSIONS

Figure 6.3 shows XRD patterns for the nanocomposites along with NCZ and PANi. The patterns were recorded at room temperature using CuKα radiation

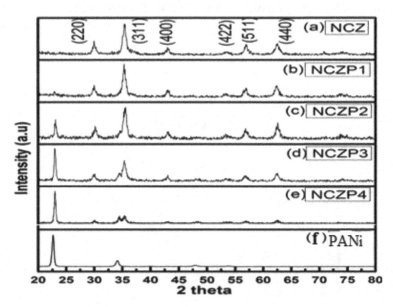

FIGURE 6.3   XRD patterns for NiCuZn ferrite(a), nanocomposites(b-e) and PANi (g).

(λ=1.5418 Å), in the 2θ range 20° to 80° with step size 0.03° operating at 40 kV and 30mA. Figure 6.2a shows the XRD pattern for as synthesized nanopowder of ferrite (NCZ) and can be clearly indexed to the seven major peaks of the spinel ferrites, which are (2 2 0), (3 1 1), (2 2 2) (4 0 0), (4 2 2), (5 1 1) and (4 4 0) planes of a cubic unit cell, which correspond to cubic spinel structure. These peaks were well matched with JCPDS card no. 03–0864. The XRD pattern for free polymer (Fig. 6.3f) does not show any sharp peaks suggesting an amorphous nature of polymer. However, the polymer displays a diffuse broad peak ranging from 15° to 30°. Figure 6.3 (b-e) gives the XRD patterns for the composites under investigation. It is observed from the figure that the degree of crystallinity of the samples increased with annealing 30h milled sample at 150°C. The XRD patterns for nanocomposites show crystalline peaks due to the presence of ferrite phase in these nanocomposites. The diffused peak of the polymer disappeared with an increase of ferrite content in the composites due to the ferrite nanoparticles interfered with polymer chains. It can also be observed from the figures that the diffraction peaks were broadened initially for pure ferrite and become narrow with higher additions of polymer. Crystallite size has been calculated using Scherer's equation (D_m = Kλ/β cos θ, K is a constant, λ wavelength of X-rays, β the full width half maximum and θ the diffraction angle), based on the widening of three strongest peaks of a certain phase. The crystallite size has been calculated using the above formula and presented in the Table 6.1. It can be observed from the table that the crystallite size of the composites is in nanorange and decreases with an increase of polymer content.

TABLE 6.1 Experimental Data on Nanocomposites

Sample	Crystallite size (nm) XRD	Particle size (nm)	Lattice parameter (A°)	Density g/cm^3	Ms (emu/g)	Hc (Oe)	Real permittivity at 1MHz
NCZ	31	30	8.4398	5.250	23.57	20	20
NCZP1	30	50	8.4231	5.236	17.6	10	17
NCZP2	26	64	8.4224	5.210	13.8	10	20
NCZP3	27	72	8.4025	5.128	8.24	8	15
NCZP4	18	83	8.4036	5.119	5.5	5	13

The morphologies of ferrite, polyaniline and its nanocomposite were examined by SEM and were given in Fig. 6.4. It can be seen from the SEM picture of ferrite nanoparticles (Fig. 6.3a) indicated a spherical shape with lots of agglomeration between ferrite particles and an average diameter of 30 nm. As observed in Fig. 6.3b, polyaniline was estimated to have an average diameter of 60 nm and a length of 100 nm. Polyaniline is tubular in shape and tends to form an entangled network that is useful for effective transport of electrons. Fig. 6.3(c and d) clearly showed that ferrite nanoparticles are assembled on the surface of polyaniline. The inset of Fig. 6.3c shows the possible way by which the ferrite nanoparticles were attached onto polyaniline surface. The enlarged image of the nanocomposite in Fig. 6.3d indicates that ferrite nanoparticles were uniformly attached on the polyaniline surface. No aggregation of ferrite was observed on the polyaniline surface.

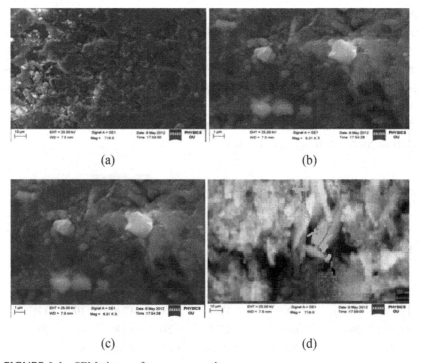

(a)   (b)

(c)   (d)

FIGURE 6.4   SEM pictures for nanocomposites.

It can be seen from the SEM micrographs of nanocomposites exhibit a two-phase system with polymer grains (white) and ferrite grains (black). Energy-dispersive X-ray spectroscopy has been used to confirm the grains of these two phases. The connectivity of NiCuZn ferrite grains is dispersed by the distribution of polymer grains, which leads to the variation of magnetic properties of the composites. The average grain sizes of the samples are estimated and results are given in Table 6.1. It can be seen from the table that the grain size of the composites is somewhat larger than the crystallite size calculated by the XRD method, which is due to the existence of the amorphous polymer matrix and the partial particle aggregation. The theoretical density of the composites is calculated from equation: $\rho = m_1 + m_2 / m_1 \rho_1 + m_2 \rho_2 + \rho_1 \rho_2$, where m_1 and m_2 are the masses of NiCuZn ferrite and polymer, respectively, ρ_1, ρ_2, and ρ the theoretical densities of NiCuZn ferrite, polymer, and composites, respectively. The bulk density of the composites is measured by Archemidics principles and they are in the range of 95%–96% of the theoretical density. The bulk density of the nanocomposites is found to increase (Table 6.1) with increasing polymer content, because NiCuZn ferrite and polymer can be densified by the interaction of the two phases that coexist in one material. From the SEM pictures, it can be observed that no mismatch has occurred in the composites, which indicates that the two phases have better co-firing properties. Table 6.1 also gives values of lattice constant (a) and porosity (%P) for the composites. It can be seen from table that the lattice constant is found to increase with an increase of ferrite phase in the composite. It is due to the fact that growth of the spinel lattice increases by the incorporation of polymer content. The average porosity in present nanocomposites is 6%.

The chemical structures of ferrite, polyaniline and nanocomposites were characterized by Fourier Transform Infrared (FT-IR) spectroscopy. Figure 6.5 shows FT-IR spectra of the ferrite (NCZ), nanocomposites (samples NCZP1, NCZP2, NCZP3, and NCZP4) of varied fractions of polyaniline with respect to ferrite, and polyaniline. For the ferrite, the band at around 588 cm^{-1} is observed in the spectrum of sample and which is attributed to the tetrahedral stretching vibration in the crystal lattices of the ferrite (Hsiang et al., 2005). The bands observed in ferrite at 1635 cm^{-1}, 1400 cm^{-1} and 1122 cm_1 are due to the vibration of the C–O, C–O–C and C–H groups of the surfactant. The characteristic peaks of polyaniline

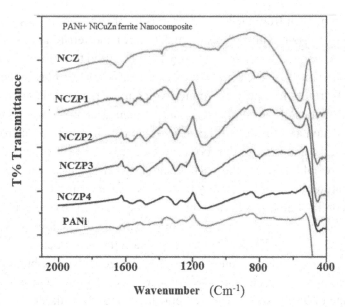

FIGURE 6.5   FTIR spectra for nanocomposites.

(sample f) are observed at 1612, 1558, 1473, 1300, 1240, 1112 and 796 cm⁻¹. The band appeared at 1612 cm⁻¹ is attributed to the C-N vibration of the quinoid ring. The bands observed in the spectrum of PANI with absorption at 1558 cm⁻¹ and 1473 cm⁻¹ are attributed to the C-C stretching vibrations of the quinoid ring and benzenoid ring, respectively. The band appeared at 1300 cm⁻¹ is the stretching of C–N bonds of the second aromatic amines. The stretching of the C–N+ polaron structure was observed at 1240 cm⁻¹. The absorption at 1112 cm⁻¹ is assigned to the protonated imine group on the polyaniline backbone chain. The band observed in the spectrum at 796 cm⁻¹ corresponds to the aromatic ring deformation and the out-of-plane vibrations of C–H bond.

For the polyaniline/ferrite nanocomposite samples, the characteristic peaks are similar. There are characteristic peaks of ferrite appearing at 561 cm⁻¹ shift compared with sample NCZ, and the characteristic peaks of polyaniline (PANi). The intensity of the peak at 561 cm⁻¹ due to the ferrite decreased with the decrease in weight percentage of ferrite in the nanocomposite. The bands appeared at 1400–1600 and 1130 cm⁻¹ of the composites show the coupling effect of ferrite and polyaniline.

Figure 6.6a shows the frequency dependence of the dielectric constant for all the samples in the range 100Hz to 1MHz at room temperature. It can be observed from figure that as frequency increases dielectric constant decreases in low frequency and remains constant at higher frequency. It can also be seen from the figure that with an addition of polymer, the dielectric increases to high value and decrease as for the higher additions of polymer. The dispersion occurring in the lower frequency region is attributed to interfacial polarization, since the electronic and atomic polarizations remain by and large unchanged at these frequencies. In the high frequency region the decrease in the value of dielectric constant is small and constant. The polarization mechanism contributing to the polarizability is observed to show lagging with the applied field at these frequencies. This can result in reduced polarization leading to diminished $\acute{\varepsilon}$ values at higher frequencies. The variation in dielectric constant can be explained in a different perspective by considering the ferrite system as a heterogeneous system with grains and grain boundaries possessing different conducting properties. Previously, the dielectric dispersion observed in a number of ferrite composite systems was explained satisfactorily on the basis of the Maxwell-Wagner theory of interfacial polarization in consonance with

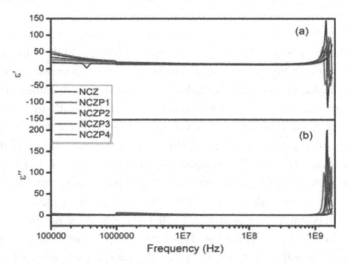

FIGURE 6.6   Frequency variation of (a) real (b) imaginary part of permittivity for nanocomposites.

the Koop's phenomenological theory. According to this model, it is the conductivity of grain boundaries that contributes more to the dielectric value at lower frequencies. In ferrites it is observed that the mechanism of dielectric polarization is similar to the mechanism of electrical conduction. The variation of dielectric constant can hence be related to the collective behavior of both types of electric charge carriers, electrons and holes. Hence, the electrons exchanging between Fe^{2+} and Fe^{3+} ions and the holes that transfer between Ni^{3+} and Ni^{2+} ions are responsible for electric conduction and dielectric polarization in these composites. At higher frequencies, the frequency of electron/hole exchange will not be able to follow the applied electric field, thus resulting in a decrease in polarization. Consequently, the dielectric constant remains small and constant.

The frequency variation of imaginary part of permittivity (ἔ) for all samples is shown in Fig. 6.6b. It can be seen from the figure that the ἔ in all samples has been found to low and remains constant and shows a resonance and antiresonance behavior has observed around 1 GHz. The reason for obtaining low values of ἔ is due to the curtailing of the Fe^{2+} ions on account of the resulting in better stoichiometry of crystal structure. The resonance peak observed in the frequency dependence of ἔ around 1GHz may be due to the hoping frequency of electrons is equal to the applied field frequency, maximum electrical energy is transferred to the oscillating ions and power loss shoots up, thereby resulting in resonance.

Figure 6.7 gives the complex permeability spectra for NiCuZn ferrite and nanocomposites. It can be seen from the figure that the value of real part of permeability (μ') is found to increases with an increase of polymer content in the composite. The value of real part of permeability (μ') for all the samples under investigation remains constant with an increase of frequency from 100kHz to 20 MHz. With further increase of frequency the value of μ' is found to increase slightly and shows domain wall dispersion at about 110 MHz. The imaginary part of permeability (μ'') decreases with an increase of frequency upto 20 MHz in pure ferrite and composites. Finally the μ'' gradually increased with an increase of frequency and took a broad maximum at a frequency of 110 MHz, where the real permeability shows domain wall dispersion. From the complex permeability spectra on nanocomposites, it is observed that the real part of permeability (μ') measured almost constant, until the frequency was raised to a certain value

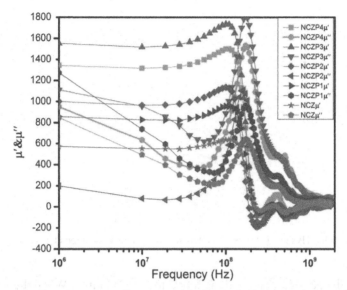

FIGURE 6.7   Complex permeability spectra for NiCuZn ferrite and nanocomposites.

and then began to decrease at higher frequency. The imaginary part of permeability (μ'') gradually deceased with the frequency and took a broad maximum at a certain frequency, where the real permeability (μ') rapidly decreases. This feature is well known domain wall resonance. The frequency dependence of permeability observed in present samples can be related to two types of magnetization mechanisms: spin rotational magnetization and domain wall motion.

Figure 6.8 shows the hysteresis loops of the pure ferrite and nanocomposites, expect for the sample NCZP5. This is because, this sample contains 90% polymer and 10% of ferrite. From the hysteresis loops the values of M_S and H_C for ferrite and composites are obtained and results are presented in the Table 6.2. It can be seen from the table that the saturation magnetization (M_S) of pure ferrite is 23.57emu/g at 300K, and decreased for the composites with an increase of polymer content. In all composites, a notably small value of coercive force is observed with negligible retentivity, which indicates the ferrimagnetic nature of the material (Xiao, et al., 2006). The room temperature saturation magnetization (Ms) of the NiCuZn ferrite decreases with an increase of polymer content in nanocomposites.

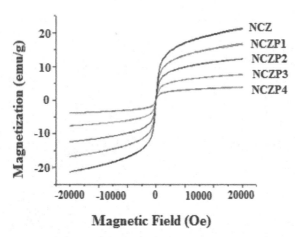

FIGURE 6.8   Hysteresis loops of the ferrite (NCZ) and nanocomposites.

Ms of the NCZP1 composite is found to be 17.6emu/g. When the volume fraction of ferrite was decreased to 30%, the Ms Value decreased drastically to 5.5 emu/g. According to the equation $M_s = \varphi m_s$, M_s is related to the volume fraction of particles (φ) and the saturation moment of a single particle (m_s). A similar observation was reported by Agarwal et al. (2011). Therefore, the saturation magnetization of the NiCuZn ferrite filled polymer nanocomposite depends mainly on the volume fraction of magnetic ferrite filler. The magnetization measurement indicated that NiCuZn ferrite-filled polymer composite could serve as a typical EMI suppressor in the frequency range of 100 kHz to 20 MHz.

The shielding effectiveness (SE) of the ferrite-on-polyaniline nanocomposite has been characterized as a ratio of the power received without shielding material (P_0) to that of the power received with shielding material (P_1). Figure 6.9 shows the shielding effectiveness of nanocomposites containing the varying weight percent of polyaniline. The shielding effectiveness increased with the increasing weight percentage of PANI in a similar way to the dielectric properties of the nanocomposites. The material can be used for the shielding purpose if the SE is higher than 5 dB and 15 dB, respectively, in the low frequency range. The present nanocomposites possess required shielding effectiveness (6 to 20 dB) in the low frequency range of 30 MHz to 200 MHz when the weight percent of polymer is above 60%. It can also be seen that the shielding effectiveness of the presently investigated nanocomposites decreased with an increase

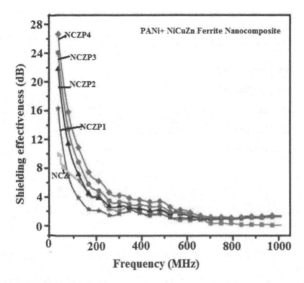

FIGURE 6.9   Shielding effectiveness (dB) for nanocomposites.

in frequency, suggesting that the ferrite-polyaniline nanocomposites are effective for the electromagnetic interference shielding at low frequencies.

6.5   CONCLUSIONS

A series nanocomposites of NiCuZn ferrite-on-polyaniline are success-fully synthesized with high dielectric absorbing properties by an in situ emulsion polymerization method. Polyaniline was added with hydrochlo-ric acid to improve its interactions with ferrite nanoparticles. FT-IR studies show that the existence of strong interfacial interactions between ferrite nanoparticles and PANi surface. From the SEM studies it is found that the ferrite nanoparticles were uniformly assembled on the surface of tubu-lar polyaniline. With the increase of the volume of NZC, the permittivity, permeability, dielectric and magnetic loss of all the composites increases. The permittivity and permeability of all the composites have shown good frequency stability and low dielectric and magnetic losses within the mea-surement range. The applicability of the present nanocomposites for the use of electromagnetic microwave absorption was examined in terms of their dielectric, magnetic properties, and shielding effectiveness.

KEYWORDS

- complex permeability
- complex permittivity
- EMI devices
- microwave-hydrothermal method
- nanocomposites

REFERENCES

1. Abbas, S. M., Dixit, A. K., Chatterjee, R., Goel, T. C., J. Magn. Magn. Mater., 2007, 309, 22.
2. Agarwal, K., Prasad, M., Sharma, R. B., Setua, D. K. 30 (2011), 329–339.
3. Baker, C., Shah, S. I., Hasanain S. K. (2004). *J. Magn. Magn. Mater.* 280, 412.
4. Bhat, N., P. Geetha, S. *Polymer Engineering and Science* (1999), 39, 1517.
5. Bins and Associates. *Plastics Additives and Compounding* (Jan. 2002), 30.
6. Brandrup, J., E. H. Immergut. *Polymer Handbook, 3rd edition*. John Wiley & Sons: New York, 1989.
7. Burke, N. A. D., Stover, H. D. H., Dawson, F. P. (2002) *Chem. Mater.* 14, 4752.
8. Chen, J. S., Poliks, M. D., Ober, C. K., Zhang, Y., Wiesner, U., Giannelis E. P. (2002) *Polymer* 43, 4895.
9. Croce, F., Appetecchi, B., Persi, L., Scrosati B (1998), *Nature* 30, 456.
10. Fang, J., Tung, L. D., Stokes, K. L., He, J., Caruntu, D., Zhou, W. L., O'Connor C J 2002 *J. Appl. Phys.* 91, 8816.
11. Flandin, L., Y. Brechet, J.-Y. Cavaille. *Composites Science and Technology* (2001), 61, 895.
12. Gass, J., Poddar, P., Almand, J., Srinath, S., Srikanth H. (2006), *Adv. Funct. Mater.* 16, 71.
13. Goldman, A. *Modern Ferrite Technology*. Van Nostrand Reinhold: New York, 1990, 84.
14. Hsiang, H. I., Chen, C. C., Tsai, J. Y. Appl. Surf. Sci. (2005), 245, 252–259.
15. Jiang, J., Li, L., Xu, F., 456 (2007), 300–304.
16. Kim, D. K., Toprak, M. S., Mikhaylova, M., Jo, Y. S., Savage, S. J., Lee, H. B., Tsakalakos, T., Muhammed M. (2004), *Solid State Phenom.* 99/100, 165.
17. Knite, M., V. Teteris, A Kiploka, J. Kaupuzs. *Sensors and Actuators A* (2004), 110, 142.
18. Koops, C. G. Phys. Rev. 83 (1951), 121.
19. Li, J. R., J. R. Xu, M. Q. Zhang, M. Z. Rong. *Carbon* (2003), 41, 2353.
20. Li, X., Han, X., Tan, Y., Xu P J. Alloys Comp. (2008), 464, 352–356.
21. Lupeiko, T. G., Lopatina, I. B., Kozyrev, I. V., Derbaremdiker, L. A. Inorg. Mater. 28 (3), (1992), 481–485.

22. Malmonge, L. F., Lopes, G. D. A., Langiano, S. D. C., Malmonge, J. A., Cordeiro J. M. M., Mattoso, L. H. C. Eur. Polym. J., 42, (2006), 3108.
23. Maxwell, J. C. Electricity and Magnetism, Oxford University Press, London, 1973.
24. Morrison, R. *Solving Interference Problems in Electronics*. John Wiley & Sons, Inc.: New York, 1995.
25. Nam, J. D., H. R. Choi, Y. S. Tak, K. J. Kim. *Sensors and Actuators A* (2003), 105, 83.
26. Okada, A., Usuki, A. (1995), *Mater. Sci. Eng.* C 3, 109.
27. Orefice, R. L., J. A. C. Discacciati, A. D. Neves, H. S. Mansur, W. C. Jansen. *Polymer Testing* (2003), 22, 77.
28. Phang SW, Tadokoro, M., Watanabe, J., Kuramoto N. (2008), Cur. Appl. Phys. 8:391–394.
29. Ponpandian, N., Balay, P., Narayanasamy, A., J. Phys. Condens. Matter. 14 (2002), 3221–3237.
30. Ramakrishna, S., J. Mayer, E. Wintermantel, K. W. Leing. *Composites Science and Technology* (2001), 61, 1189–1224.
31. Rogers, S. S., L. Mandelkern. *Journal of Physical Chemistry* (1957), 61, 945.
32. Ruiz-Hitzky, E., Aranda P (1990), *Adv. Mater.* 2 545.
33. Supriyatno, H., K. Nakagawa, Y. Sadaoka. *Sensors and Actuators B* (2001), 76, 36.
34. Tareev, B. Physics of Dielectric Materials, MIR Publishers, Moscow, 1975.
35. Vaia, R. A., Ishii, H., Giannelis E P (1993), *Chem. Mater.* 5 1694.
36. Wagner, K. W. Ann. Phys. 40 (1993), 818.
37. Wang, M. *Biomaterials* (2003), 24, 2133.
38. White, D. *A Handbook on Electromagnetic Shielding Materials and Performance*. Don White Consultants, Inc.: Germantown, MD, 1975.
39. Wilson, J. L., Poddar, P., Frey, N. A., Srikanth, H., Mohomed, K., Harmon, J. P., Kotha, S., Wachsmuth J. (2004), *J. Appl. Phys.* 95, 1439.
40. Xiao H. M., Liu X. M., Fu S. Y. (2006), Comp. Sci. Tech. 66, 2003–2008.
41. Xiao, H. M., Liu, X. M., Fu, S. Y. 66 (2006), 2003–2008.
42. Xuan, S., Wang, Y.-X. J., Yu, J. C., Leung, K. C.-F. Langmuir, (2009), 25, 11835.
43. Yang, C., Li, H., Xiong, D., Cao Z. (2009), 69, 137–144.
44. Zhang, L., Li Z. (2009), J. Alloys Comp. 469, 422–426.
45. Zhou, X. D., Gu, H. C. (2002). *J. Mater. Sci. Lett.* 21, 577.

CHAPTER 7

MACRO LEVEL INVESTIGATION ON THICKNESS VARIATION OF ROT MOULDED LLDPE PRODUCT

P. L. RAMKUMAR, DHANANJAY M. KULKARNI, and
SACHIN D. WAIGAONKAR

*BITS PILANI, K. K. Birla Goa campus, Goa–403726, India; Mobile:
+919823256780; Email: ramkumarpl@goa.bits-pilani.ac.in,
plramkumarno1@gmail.com*

CONTENTS

7.1   Introduction... 121
7.2   Materials and Methods.. 123
7.3   Results and Discussion 126
7.4   Conclusion 132
Authors' Contributions... 133
Keywords 133
References.. 133

7.1   INTRODUCTION

Rotational moulding is a currently the fastest growing sector of plastic processing industry. In rotational moulding process, preweighted quantity of plastic powder is loaded in to the mould. The mould is then bolted in to the oven where it undergoes a biaxial rotation along with the heating of mould. After certain temperature the thermoplastic powder sticks to the wall of the mould. The process is continued till the required thickness of the plastic part is obtained. After getting the required thickness, mould is cooled and demoulded (Antonio and Alfonso, 2004; Bharat et al., 2001; Crawford, 2002; Brent strong, 2006). The process is economical as several components can be produced in one piece, thereby eliminating the need of expensive fabrication. As the products are inherently stress free, components with good mechanical and service properties can be obtained. The products obtained from rotational moulding find extensive applications in various fields like agriculture, storage tanks, industrial equipment, medical devices, material handling, road/highways, automobiles, etc.

An important contribution in the development of rotational moulding technology has been made by Crawford. Crawford (2002, 1994), and Spence and Crawford, (1996) mainly concentrated on different material processibility and its structural competence for rotational moulding process. He has identified polyethylene especially LLDPE as the suitable processing material for rotational moulding process because of its wide processing window. A new approach called multi attribute decision making (MADM) method was introduced for resin selection in rotational moulding process by Sachin et al. (2008). The procedure has been adapted to rank different resins, to quantitatively assist a rotomolder to select a proper resin from a long list for a specific application.

However, a proper control over the process variable is required to obtain a quality product. The two important process variables that affect the thickness of the rot-moulded product are:

- Speed Ratio (X_1)
- Cycle Time (X_2)

A number of studies have been carried out to study the effect of these variables on mechanical properties and quality of rotationally moulded products.

Crawford (1996) identified the PIAT of LLDPE as 200–220°C to get optimal mechanical properties of the rotationally moulded product and discussed the relationship between the oven temperature and PIAT. He also provided several guidelines for reduction in cycle times, use of different materials, their properties and processing windows as well as the design aspects of rotationally moulded products (Crawford, 2002).

Recently Aissa et al. (2012) have carried out image analysis technique to determine the effect of polymer powder particle size and distribution in a biaxialy rotating spherical mould. They have developed a technique viz. the gray-level co-occurrence matrix (GLCM) to determine the mixing homogeneity as a function of time. The image analysis was used to follow the motion of polymer powders inside a spherical mould.

Tan et al. (2011) have reviewed the different methods of internal cooling of mould to reduce the cycle times in rotational moulding. The methods of cooling studied by the author include introduction of compressed air inside the mould, introduction of CO_2 at cryogenic conditions and water spray cooling. According to the authors' the water-cooling could be an effective alternative to reduce the cycle times, better structural uniformity of the plastic and hence better mechanical properties. Abdullah et al. (2007) have proposed a combination of different conditions like surface enhanced moulds, higher oven flow rates, internal mold pressure, and water spray cooling for cycle time reduction in rotational moulding. They have reported a cycle time reductions of up to 70%.

Lopez et al. (2012) have used obtained agave fiber/linear medium density polyethylene composites using rotational moulding. Their findings revealed that addition of such fibers gave better flexural and tensile properties of the product. However, the optimal fiber concentration was suggested around 10% due to inadequate wetting of the fibers and poor aesthetics with increased fiber concentration.

Based on the literature survey, it was found that the quality improvement methods like Six Sigma have not been discussed sufficiently for rotational moulding applications. This work presents application of Six Sigma methodology to study the thickness variation of rotationally moulded products for various speed ratio and cycle time. The problem of producing rotationally moulded products of specified thickness was defined, and investigations were carried out by planning and performing trials based on design of experiments (DOE). In this study, DOE approaches viz. full factorial methods were used for data analysis.

7.2 MATERIALS AND METHODS

7.2.1 MATERIAL

The polymer used was linear low density polyethylene of grade R350 A 42 having melt flow index of 4.2 g/10 min and density was 935 Kg/cm^3 supplied by GAIL India limited. The above-mentioned grade is recommended for manufacturing of water storage tanks, automobile parts, and boats, etc.

7.2.2 EQUIPMENT USED

Rotationally moulded products were produced on a lab scale electrically heated biaxial rotational moulding machine as shown in Figs. 7.1(a) and 7.1(b) using a stainless steel hollow mould having the square cross-section with polished internal surface as shown in Fig. 7.1(c). Powder weight of 250 g was

FIGURE 7.1 (a) Biaxial rotational moulding machine; (b) major and minor axis rotation; (c) stainless steel hallow mould.

used produce rotational moulding product. In all trials the oven temperature was set at 240°C. This temperature is sufficient to get a PIAT of 200°C as evident from the static/uniaxial trials taken before plan of experiments.

7.2.3 EXPERIMENTAL PROCEDURE

The internal mould surface of the mild steel mould was coated with a silicon oil based mould release agent. The process oven temperature was set at 240°C as stated above. Pre weighted quantity of 250 g of powder was placed inside the mould. The mould rotation speed was set (which varies for different trials). The mould was then bolted in to the oven where it underwent a biaxial rotation, which makes the powder to spread in the internal surface of the mould. After certain temperature the thermoplastic powder melts and sticks to the wall of the mould. The process is continued till the required thickness of the plastic part is obtained.

7.2.4 PLAN OF EXPERIMENTS

Initially, a few preliminary experiments were conducted to study the variation of thickness at random level of process variables. LLDPE used in this study had a melting temperature of 130°C while its degradation temperature was 200°C. During preliminary runs, it was observed that, processing at values of cycle time below 32 mins resulted in undercooked part owing to incomplete sintering of the polymer. On the contrary, for values higher than 42 mins polymer degradation was noticed. It was also observed that speed ratio below 3:1 resulted in non uniformity in thickness of the products. Very high speed (more than 5:1), on the other hand produced large number of internal bubbles, which intern reduces the strength of the product. Thus three levels of each process variable, coded as $(-1, 0, +1)$, were identified and are shown in Table 7.1.

As there are two variables each at two levels with three replication of each and three center points, full factorial design will yield 15 ($2^2 \times 3 + 3$) experiments. The factorial points in this scheme $(-1, +1)$ serves as preliminary stage to obtain a first order model while center points $(0, 0)$ helps to build a second order model to estimate the curvature effects.

TABLE 7.1   Process Variables and Their Levels For Rotational Moulding

Process Variable	Unit	Designation	Lowest level (−1)	Highest level (+1)
Speed ratio	Rpm	X1	3:1	5:1
Cycle time	Minutes	X2	32	42

The selected scheme of experimental design thus had fewer design points (15 trials), still having provision to yield efficient estimation of thickness (response).

7.3   RESULTS AND DISCUSSION

7.3.1   STATISTICAL ANALYSIS

The plan of experiments and Thickness values (in mm) for the different experiments are as shown in Table 7.2. To determine the significant factors

TABLE 7.2   Plan of Experiments and Corresponding Value of Thickness

Run order	Blocks	X_1 (rpm)	X_2 (mins)	Thickness(mm)
1	1	3	42	2.98
2	1	5	32	2.90
3	1	5	42	3.00
4	1	3	42	2.96
5	1	3	32	2.40
6	1	3	32	2.20
7	1	5	32	2.90
8	1	3	32	2.12
9	1	4	37	3.00
10	1	5	42	3.00
11	1	4	37	2.70
12	1	5	32	3.00
13	1	4	37	2.80
14	1	5	42	2.98
15	1	3	42	3.12

or interactions, Analysis of variance (ANOVA) was carried out at 5% level of significance ($p = 0.05$). A 'p' value less than the 0.05 can be considered as significant. This is because, at 95% level of confidence, it rejects the null hypotheses that the factors have no effect on thickness (against an alternative hypotheses that the factors have significant influence on thickness).

Referring to the ANOVA along with the 't' test results in Table 7.3, it could be seen that there were significant linear effect of process variables on the thickness. The coefficients of the regression model $|\beta|$, which in general can be written as shown in Eq. (1).

$$y = X \beta + \varepsilon$$

where,

$$[y] = \begin{bmatrix} y_1 \\ y_1 \\ \cdot \\ \cdot \\ y_n \end{bmatrix} \quad [X] = \begin{bmatrix} 1 & x_{11} & x_{12} & .. & x_{1k} \\ 1 & x_{21} & x_{22} & & x_{2k} \\ \cdot & \cdot & \cdot & .. & \cdot \\ \cdot & \cdot & \cdot & .. & \cdot \\ \cdot & \cdot & \cdot & .. & \cdot \\ 1 & x_{n1} & x_{n2} & .. & x_{nk} \end{bmatrix} \quad [\beta] = \begin{bmatrix} \beta_1 \\ \beta_2 \\ \cdot \\ \cdot \\ \beta_n \end{bmatrix} \quad [\varepsilon] = \begin{bmatrix} \varepsilon_1 \\ \varepsilon_2 \\ \cdot \\ \cdot \\ \varepsilon_n \end{bmatrix} \quad (1)$$

The procedures of obtaining these constants have been discussed at sufficiently length by Montgomery (2004). The empirical model for thickness in terms of coded units considering only significant terms from ANOVA was obtained as:

TABLE 7.3   Results of ANOVA For Thickness

Term	Effect	Coef	SE Coef	T	p
Constant		2.7967	0.03033	92.20	0.000*
X1	0.3333	0.1667	0.03033	5.49	0.000*
X2	0.4200	0.2100	0.03033	6.92	0.000*
X1*X2	−0.3600	−0.1800	0.03033	−5.93	0.000*
Ct Pt		0.0367	0.06782	0.54	0.601**

Note: *Significant at 5% level of significance.

**Non-significant at 5% level of significance.

$$\text{Thickness} = -4.75 + 1.49\,X_1 + 0.18\,X_2 - 0.03\,X_1\,X_2 \qquad (2)$$

Equation (2) can be conveniently used to predict the thickness for any combination of process variables within the regime of experimentation (i.e., −1 to +1).

The adequacy of the model was checked with the help of a normal probability plot between the residuals and the predicted values. The residual is the difference between the observed and the predicted value from the regression. If the points of the plot are seen closer to the straight line, then the data is normally distributed. The normal probability plot and the residuals vs. fits for the thicknesses are shown in Figs. 7.2 and 7.3, respectively. Figure 7.2 reveals that residuals fall on a straight line and the errors are distributed normally. Figure 7.3 reveals that most of the residuals fall closer to the zero residual line, therefore giving minimum error in fitting regression equation. From these two plots also it could be concluded that the model proposed was adequate.

7.3.2 INTERACTION PLOT

To identifying the interactions effect of all the parameters on thickness, the interaction plot was generated using Minitab 14.0. The interaction plot was shown in Fig. 7.4. In this plot, the variation of thickness with

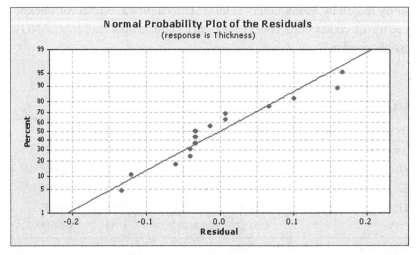

FIGURE 7.2 Normal Probability plot for thickness.

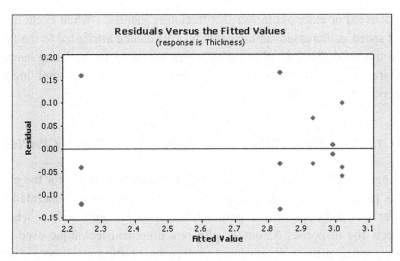

FIGURE 7.3   Residual plot for thickness.

respect to the combination of process parameters at different levels is represented. The intersecting lines show that when both the parameters are varied simultaneously, the combined effect will be evident on the

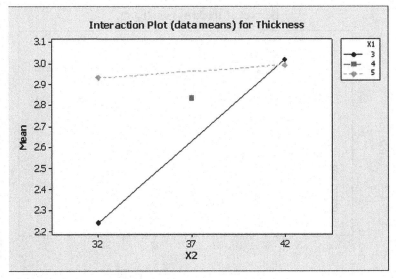

FIGURE 7.4   Interaction plot for thickness.

response. For example in Fig. 7.5 thickness increases when cycle time and speed is increased simultaneously. This can be attributed to the fact that the build up of the plastic on the mould acts as an insulating barrier separating the external heating from the resin that is still tumbling in interior of the mould.

7.3.3 CONTOUR PLOTS AND RESPONSE SURFACE GRAPHS

Using Minitab 14.0 software, contour plots and 3D graphs for the process parameters with respect to thickness variation were generated in order to find the responsible parameters or combinations of these which affects the response. A contour plot is a graphical technique used for representing a three-dimensional surface by plotting constant z slices called contours on a 2-dimensional format. A surface graph provides a three dimensional view, which gives a clear picture of the response surface. The contour and surface plot of the response was shown in Figs. 7.6 and 7.7. As observed from the plots the thickness increases

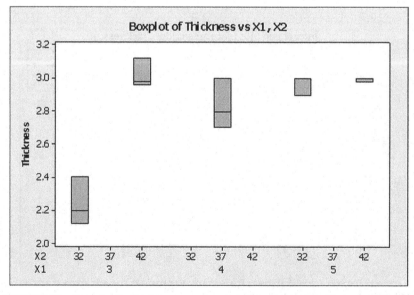

FIGURE 7.5   Box plot for thickness.

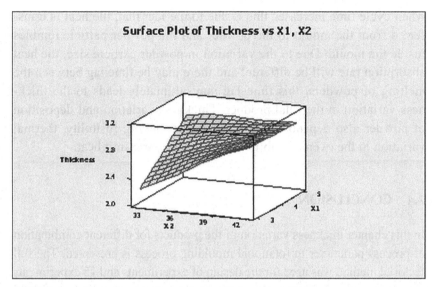

FIGURE 7.6 3D Surface plot for cycle time and speed vs. thickness.

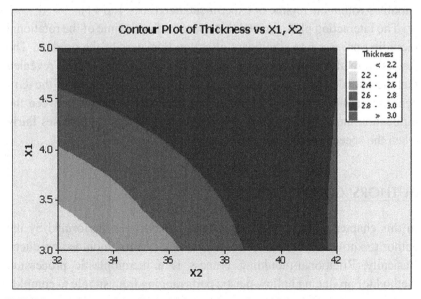

FIGURE 7.7 Contour plot for cycle time and speed vs. thickness.

when cycle time increases this is due to the fact that, the heat is transferred from the mould to the powder and the powder particle tumbles inside the mould. Due to the variation in powder particle size, the heat absorption rate will be different and there may be time lag between the melting of powders, this time lag may ultimately leads to the thickness variation in the final product. Thickness variation and deposition of powder also depends upon particle distribution, fusibility, thermal variation in the oven and ability of the resin to transmit heat.

7.4 CONCLUSION

In this chapter thickness variation of the products for different combination of process parameter in rotational moulding process is presented. The full factorial method was used for the design of experiments and 15 experiments were designed and conducted. Experimental result using ANOVA confirmed that both the process parameters, that is, cycle time and speed has significant effect on thickness of the rot-moulded product. The result obtained confirms that there is a linear relation between the process parameters and response. Using Eq. (2) the thickness can be predicted for any combination of process variables within the regime of experimentation (i.e., −1 to +1).

The interaction plots showed that speed and cycle time of the rotational moulding process had significant effect on thickness of the product. The results were also analyzed using surface and contour plot which revealed that the thickness increases when cycle time increases because of the temperature variation inside the mould and thickness variation between the replication increases at very low speed ratio (3:1) and it decreases fairly when the speed ratio is high (5:1).

AUTHORS' CONTRIBUTIONS

In this chapter a macro level investigation have been performed by the author to know the effect of process parameter on rotomoulded products. Basically, Rotational-moulding process is a thermoplastic processing method for producing hollow plastic parts ranging from simple to complex geometries. Out of all polymers, polyethylene is preferred for rotational moulding process because of its favorable processing window. Effect of thickness for various speed ratios and cycle time is the biggest challenge in this process. Moulders have to depend heavily upon trial and error methods as well as experience of the operator to predict the thickness for a particular speed ratios and cycle time. In this work, an attempt has been made to investigate the thickness of the rotational moulded parts for different speed ratios and cycle time using experiments and statistical techniques. Experimental runs and analysis based on design of experiments (DOE) revealed that thickness of the part is severely affected by both of these process parameters. The procedure would quantitatively assist the rotomoulders to select a proper speed ratio and cycle time for required thickness of the rotational moulded product.

KEYWORDS

- **design of experiment**
- **LLDPE**
- **process parameter**
- **rotational moulding**

REFERENCES

1. Abdullah, M. Z., Bickerton, S., Bhattacharyya, D. Rotational Moulding Cycle Time Reduction Through Surface Enhanced Molds: Part A—Theoretical Study. *J. Polym. Eng. Sci.* 47, 1406–1419 (2007).
2. Amara, A., Carl, D., Denis, R. Characterization of Polymer Powder Motion in a Spherical Mold in Biaxial Rotation. *J. Polym. Eng. Sci.* 52, 954–963 (2012).
3. Antonio, G., Alfonso, M. Powder-shape analysis and sintering behavior of high-density polyethylene powders for rotational moulding. *J. Appl. Polym. Sci.* 92, 449–460 (2004).
4. Bharat, I. C., Elizabeth T., John, V. Processing enhancers for rotational moulding of polyethylene. *J. Polym. Eng. Sci.* 41, 1731–1742 (2001).
5. Brent strong, A. *Plastics Materials and Processing*, 3rd ed., Pearson Education Inc: New Jersey, (2006).
6. Crawford, R. J. Causes and Cures of Problems during Rotomoulding. *J. Rotation.* 3, 10–14 (1994).

7. Crawford, R. J. Recent Advances in the Manufacture of Plastic Products by Roto-moulding. *J. Mater. Process. Technol.* 56, 263–271 (1996).
8. Crawford, R. J., James, L. T. Rotational Moulding Technology, William Andrew Publishing: New York (2002).
9. Lopez-Banuelos, R. H., Moscoso, F. J., Ortega-Gudin, P., Mendizabal, E., Rodrigue D., Gonzalez-Nunez, R. Rotational Moulding of Polyethylene Composites Based on Agave Fibers. *J. Polym. Eng. Sci.* 52, 2489–2497 (2012).
10. Montgomery, D. *Design and Analysis of Experiments*, 5th ed., John Wiley and sons. Inc: New York, (2004).
11. Spence, A. G., Crawford, R. J. The Effect of Processing Variables on the Formation and Removal of Bubbles in Rotationally Moulded Products. *J. Polym. Eng. Sci.* 36, 993–1009 (1996).
12. Tan, S. B., Hornsby, P. R., McAfee, M. B., Kearns M. P., McCourt, M. P. Internal Cooling in Rotational Moulding— A Review. *J. Polym. Eng. Sci.* 51, 1683–1692 (2011).
13. Waigoankar S., Babu, B. J. C., Prabhakaran, R. T. D. A New Approach for Resin Selection in Rotational Moulding. *J. Reinf. Plast. Compos.*, 27, 1021–1037 (2008).

CHAPTER 8

MOLECULAR STRUCTURE AND PROPERTY RELATIONSHIP OF COMMERCIAL BIAXIALLY ORIENTED POLYPROPYLENE (BOPP) BY DSC, GPC AND NMR SPECTROSCOPY TECHNIQUES

RAVINDRA KUMAR, VEENA BANSAL, S. MONDAL, NITU SINGH, A. YADAV, G. S. KAPUR, M. B. PATEL, and SHASHIKANT

Indian Oil Corporation Limited, Research and Development Centre, Sector-13, Faridabad-121007, India; E-mail: kumarr88@indianoil.in

CONTENTS

Abstract 136
8.1 Introduction 136
8.2 Experimental 137
8.3 Results and Discussion 138
8.4 Conclusions 141
Acknowledgement 142
Keywords 142
References 142

ABSTRACT

Polypropylene (PP) is one of the most important polymers in use. The one important application of PP is the production of biaxially oriented PP film (BOPP), which finds major applications in the area of packaging. The principal structural factors affecting end use properties of PP are microstructure (i.e., tacticity distribution, other stereo-defects and presence/absence of any co-monomer), molecular weight, molecular weight distribution (MWD) and crystallinity.

In the present work, five commercial BOPP resins manufactured using different process technologies and catalyst systems have been characterized in depth using nuclear magnetic resonance (NMR) spectroscopy, gel permeation chromatography (GPC) and differential scanning calorimetry (DSC) techniques. The resins were separated into xylene soluble (XS) and xylene insoluble (XIS) fractions. Each of the XS and XIS fractions were then subjected to high temperature ^{13}C-NMR spectroscopy measurement in order to determine their detailed microstructure (i.e., tacticity distribution, %mmmm pentad distribution, other stereo-defects and presence/absence of any co-monomer). High temperature gel permeation chromatography (HTGPC) of these fractions was carried out in order to determine their molecular weights and molecular weight distribution (MWD). Differential scanning calorimetry (DSC) studies were carried out to determine the melting temperature (Tm), crystallization temperatures (Tc) and crystallinity of the resin samples. The analysis showed that these resins not only had differences in the content of xylene soluble fractions but also had distinct differences in the content of %mmmm pentad contents in the XIS fractions. The % mmmm pentad content in the XIS fractions have been found to vary from 88 to 92% and the stereo defect varied from 1.63 mol% to 2.17 mol% in the different samples. The above differences in the microstructural parameters arising due to the use of different process technology and catalysts have been correlated with the processability of these BOPP resins.

8.1 INTRODUCTION

Biaxially Oriented Polypropylene (BOPP) film is a thermoplastic polymer film of polypropylene (PP) with an orderly molecular structure formed by biaxial orientation process. The processes improve the optical and gas barrier properties of the film. BOPP film is known for its excellent clarity, high tensile and impact strength, good dimensional stability and flatness, low electrostatic charge, waterproof and moisture repellent, excellent transparency, lower density, gas and moisture barrier properties. The principal structural factors affecting end use properties of PP are microstructure (i.e., tacticity distribution, other stereo-defects and presence/absence of any co-monomer), molecular weight, molecular weight distribution (MWD) and crystallinity.

In the present work, molecular level characterization of commercial BOPP resins manufactured using different process technologies and catalyst systems have been carried out using nuclear magnetic resonance (NMR) spectroscopy, gel permeation chromatography (GPC) and differential scanning calorimetry (DSC) techniques for the basic understanding of structure-property relationship. The resins were separated into xylene soluble (XS) and xylene insoluble (XIS) fractions. The influence of the type of catalyst and process on microstructures of industrial Ziegler Natta (ZN) PP has been classically studied by fractionation and 13C NMR analysis of the fractions (Paukkeri et al., 1993; Morini et al., 1996). However, the generality of the distribution of defects has been hampered because, historically, ZN catalysts have been modified to increase the isotacticity level. The concentration of stereo defects, regio defects, or comonomers randomly inserted in the PP chain can now be controlled by proper selection of catalyst. Thus, properties of iPPs can be studied as a function of the type and the fractional content of these defects for a fixed molecular weight and molecular weight distribution. As the concentration of defects in industrial iPPs will be increased as a mean to tailor made properties, such as reduce crystallinity, lower the softening point improve impact properties and processability.

8.2 EXPERIMENTAL

Five commercial BOPP resins manufactured using different process technologies and Ziegler Natta catalyst systems have been used in this study, which details are given in Table 8.1. Details of BOPP resins are given in Table 8.1.

TABLE 8.1 MFR and Tensile Properties of BOPP Resins

Property	PP-1	PP-2	PP-3	PP-4	PP-5	PP-6
Melt Flow Rate (g/10 min) (230°C/2.16 kg)	3.0	2.9	3.2	3.4	3.0S	3.1
Tensile strength at yield (MPa)	34	34	35	36	34	34
Tensile elongation at yield (MPa)	11	9	9	12	11%	11

8.2.1 NMR

All NMR spectra were generated on Bruker 500 MHz NMR spectrometer using 10 mm dual probe. The samples were prepared and measured according to the procedure described in the literature (Randall, 1976). Peak assignment was done by using the results described in reference (Busico et al., 1997).

8.2.2 DSC

Crystallization was studied by DSC. The DSC (TA Instrument 2920) was calibrated using indium before use. The sample chamber was kept under a constant flux of nitrogen. All DSC melting scans were recorded at a heating rate value of 10°C/min.

8.2.3 GPC

HTGPC analysis was carried out on PLGPC 220 equipment fitted with three detectors (Refractive Index, Viscosity and light scattering detector).

8.3 RESULTS AND DISCUSSION

The resins were separated into XS and XIS fractions according ASTM D5492. Each of the XS and XIS fraction were then subjected to high temperature ^{13}C-NMR spectroscopy measurement in order to determine their detailed microstructure (i.e., tacticity distribution, %mmmm pentad distribution, other stereo-defects and presence/absence of any co-monomer). NMR spectrum (methyl region) of PP-XIS-6 XIS fraction shown in Fig. 8.1.

The pentad sequence distribution of XIS fractions and the concentrations of stereo defects, obtained from the 13C NMR spectra as half the content of mmmr pentads are listed in the Table 8.2. The analysis showed that these resins not only had differences in the content of xylene soluble fractions (Table 8.2) but also had distinct differences in the content of %mmmm pentad contents in the XIS fractions. The %mmmm pentad content in the XIS

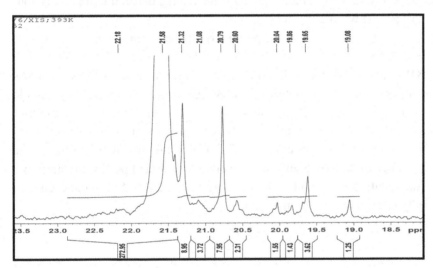

FIGURE 8.1 Methyl region of PP-XIS-6 NMR spectrum.

TABLE 8.2 Stereo Defects and Sequence Distribution in Different BOPP Resins

Sequence distribution	PP-XIS-1	PP-XIS-2	PP-XIS-3	PP-XIS-4	PP-XIS-5	PP-XIS-6
mmmm	88.3	90.2	89.1	89.7	92.1	90.2
mmmr	4.34	3.96	4.34	4.28	3.26	3.28
rmmr	1.3	1.1	1.0	1.1	0.9	1.2
mmrr	2.7	1.9	2.8	2.9	2.0	2.6
Mmrm+rmrr	1.2	1.8	0.9	1.0	0.7	0.8
rmrm	0.0	0.0	0.7	0.6	0.0	0.5
rrrr	0.9	0.3	0.8	0.5	0.5	0.5
rrrm	0.5	0.3	0.6	0.5	0.3	0.5
mrrm	1.3	0.9	1.2	1.3	1.0	1.2
Defects Mol %	2.17	1.98	2.17	2.14	1.63	1.64

fractions have been found to vary from 88 to 92% with the stereo defect varying from 1.63 to 2.17 mol% in the different samples (Table 8.2).

Defects are irregularities that depress melting and slow crystallization of polymers. Therefore, XIS fractions with high defect content are expected to display lower melting and crystallization temperatures. The observed melting and crystallization temperatures are generally higher for XIS fractions of lower defect contents, as expected from the fractionation method. The NMR data clearly indicate that the defect distribution is non-uniform between XIS fraction.

The high temperature gel permeation chromatograph profile (Fig. 8.2) with labeling as RK-7 to RK 12 (for PP-XIS-1 to PP-XIS-6) belongs to XIS fractions of BOPP. The molecular weight distribution is more or less similar with little variations. The comparative GPC profile for sample no PP-XIS-1 to PP-XIS-6 indicates that that the sample no. PP-XIS-1 (RK-7), besides containing the polymeric polyolefin species also contains low molecular weight components. One of the probable reasons of the presence of this lower molecular weight component peak is the improper fractionation of xylene solubles /insolubles. PP-XIS-2 sample contain ethylene content in small amount.

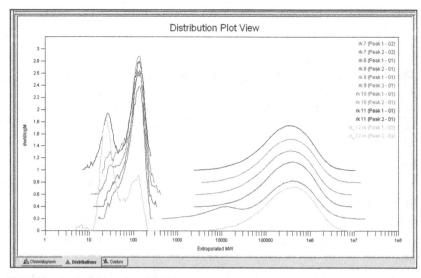

FIGURE 8.2　MWD curve of XIS fractions.

TABLE 8.3 Melting Temperature, Crystallization Temperature, Crystallinity, MWD and Stereo Defects in Xylene Insoluble Fractions of XIS

	PP-XIS-1	PP-XIS-2	PP-XIS-3	PP-XIS-4	PP-XIS-5	PP-XIS-6
% XS	3.8	3.9	5.1s	4.4	3.7	3.7
Tm (°C)	163.8	165.6	163.9	164.0	165.9	167.0
Tc (°C)	117.0	117.3	117.1	118.4	117.9	117.9
% Xc	46.9	45.5	43.2	44.5	47.4	47.8
Mw	621000	542000	514000	552000	500000	582000
Mn	264000	122000	113000	122000	264000	151000
Mw/Mn	3.3	4.4	4.5	4.5	4.6	3.8

In analyzing the molar mass data of Table 8.3 we observe that there is no systematic trend in molecular weight with increasing (mol % defect) for these set of XIS fractions.

Xylene soluble fractions were also estimated according ASTM method and analyzed by 1H and 13C NMR spectroscopy. Xylene solubles have been found to vary from 3.7 to 5.1%. The melting and crystallization temperature and the overall crystallinity decrease with increasing amounts of XS. The isotacticity (%mmmm) of XIS fraction was considered to be the same within the limit of experimental error for all the polymers employed. Therefore, the overall crystallinity is affected mostly by the amounts of XS. The number average meso sequence length was determined from the measured sequence distributions. In general, the longer the meso run, the higher the material crystallizability. Pentad sequence in PP-XS-1 xylene soluble also estimated and found 14.6 % isotactic sequence (%mmmm) and 15 % rmrm sequence. The pentad of rmrm existing in the polymer may improve the flexibility of the polymer.

8.4 CONCLUSIONS

In conclusion, we may state that BOPP films are thermoplastic polymer of polypropylene with an orderly molecular structure formed by biaxial orientation process that provides us with a unique opportunity to study the microstructure (i.e., tacticity distribution, other stereo-defects and presence/absence of any co-monomer), molecular weight, molecular weight distribution (MWD), crystallinity and influence of the process and catalyst on their properties. The study provides much help for basic understanding of molecular level structure and it has been found that there is no major difference present w.r.t. structure and physical properties of all commercially available BOPP resins in Indian market manufactured by different process.

ACKNOWLEDGEMENT

The authors are thankful to the management of IOC, R&D Faridabad, for allowing publishing present work.

KEYWORDS

- **BOPP resins**
- **differential scanning calorimetry**
- **gel permeation chromatography**
- **molecular weight distribution**
- **polypropylene**
- **tacticity distribution**

REFERENCES

1. Busico, V., Cipullo, R., Monaco, G., Vacatello, M. (1997), *Macromolecules,* 30, 6251.
2. Morini, G., Albizzati, E., Balbontin, G., Mingizzi, I., Sacchi, M. C., Forlini, F., Tritto, I. (1996), Macromolecules, 29, 5770.
3. Paukkeri, R., Vaananen, T., Lehtinen, A. (1993), Polymer, 34, 2488.
4. Randall, J. C. (1976), *J. Polym. Sci.* 14, 2083.

CHAPTER 9

LEAD AND CADMIUM ION REMOVAL BY NOVEL INTERPENETRATING POLYMER-CERAMIC NANOCOMPOSITE

K. SANGEETHA, and E. K. GIRIJA

Department of Physics, Periyar University, Salem–636 011, India; Tel.: +91 9444391733; Fax: +91 427 2345124; E-mail: girijaeaswaradas@gmail.com

CONTENTS

Abstract.. 143
9.1   Introduction.. 144
9.2   Experimental.. 145
9.3   Results and Discussion 148
9.4   Conclusions.. 151
Acknowledgment.. 151
Keywords 152
References.. 152

ABSTRACT

Release of heavy metals into environment through industrial effluents may pose a serious threat due to the growing discharge, toxicity and other adverse effects. In this report, sorption of Pb^{2+} and Cd^{2+} by a bionanocomposite prepared by incorporating nano-sized hydroxyapatite into alginate-gelatin polymeric matrix has been discussed. The adsorption kinetics and equilibrium studies revealed that the adsorption process followed a Pseudo-second-order kinetic and Langmuir isotherm model respectively. This study shows that the hydroxyapatite-alginate-gelatin composite holds great potential to remove the cationic heavy metals from the environmental wastewater.

9.1 INTRODUCTION

Metallic species mobilized and released into the environment by the technological activities of humans tend to persist indefinitely, circulating and eventually accumulating throughout the food chain, thus posing a serious threat to the environment, animals, and humans. During the last decade extensive attention has been paid on the management of environmental pollution caused by hazardous material such as heavy metal, textile dye and other organic and inorganic wastes. Decontamination of heavy metals in the soil and water around industrial plants has been a challenge for a long time. A large number of methods have been developed for the removal of heavy metals from liquid wastes such as precipitation, ion exchange, evaporation, membrane processes, electroplating, etc. However these methods have several drawbacks such as unpredictable metal ion removal, high reagent requirement, generation of toxic sludge, etc. Adsorption is a cost effective and exceptional tool for removing heavy metal ion from the aqueous solution.

Nowadays, technology is devoted to the development of novel nontoxic adsorbents, which are capable of satisfying special requirements in terms of immobilization of heavy metal ions from aqueous media. Hydroxyapatite (HA) is the inorganic component of bones and teeth which has been widely utilized in the purification and separation of biological macromolecules such as proteins, enzymes, viruses and nucleic acids since 1956, the adsorption of protein to HA occurs in cationic and anionic modes, depending on the operating buffer conditions (Tiselius et al., 1956; Schröder et al., 2003; Girija et al., 2012). It has been used as a cation exchanger and has a very high capacity for removing divalent heavy metal ions from contaminated water (Sundaram et al., 2008; Xu et al., 1994).

According to the Hard–Soft Acid Base (HSAB) Principle, hard ions which bind F^- strongly, such as Na^+, Ca^{2+}, Mg^{2+} could form stable bonds with OH^-, HPO_4^{2-}, CO_3^{2+}, R–COO$^-$ and =C=O, which are oxygen-containing ligands. Contrast to hard ions, soft ions such as Hg^{2+} and Pb^{2+} form strong bond with CN^-, R–S$^-$, –SH$^-$, NH^{2-} and imidazol, which are groups containing nitrogen and sulfur atoms. Borderline or intermediate metal ions such as Zn^{2+} and CO_2^+ are less toxic. Hard ions mainly show ionic nature of binding, whereas soft ions binding exhibit a more covalent degree. Some active sites involved in the metal uptake are determined by using techniques of titration, infra-red and Raman spectroscopy, electron dispersive spectroscopy (EDS), X-ray photoelectron spectroscopy (XPS), electron microscopy (scanning and/or transmission), nuclear magnetic resonance (NMR), X-ray diffraction analysis (XRD).

Furthermore nano sized hydroxyapatite have been considered as an ideal low cost adsorbent due to its least solubility and high sorption capacity for metal ions, however it is usually obtained by calcination. Calcination leads to a significant decrease in the specific surface area by particle coalescence and densification (Bailliez et al., 2004). To improve its applicability for the detoxification of contaminated water, naturally occurring polymers such as alginate which is rich in functional groups can be used as matrix for the immobilization of HA powders. Recently we reported the effect of gelatin on the in situ formation of alginate/hydroxyapatite nanocomposite. Presence of gelatin influenced the crystallinity of apatite formed and also the microstructure of the resulting composite (Sangeetha et al., 2013). Here we report the lead and cadmium removal efficiency of nHA immobilized on alginate-gelatin interpenetrating polymeric matrix.

9.2 EXPERIMENTAL

9.2.1 MATERIALS AND METHODS

Calcium nitrate tetrahydrate [$Ca(NO_3)_2.4H_2O$, 98%], di ammonium hydrogen phosphate [$(NH_4)_2HPO_4$, 99%], gelatin and ammonia solution [NH_4OH, 25%] were obtained from Merck. Sodium alginate and lead nitrate [$Pb(NO_3)_2$] were obtained from Loba chemie (India) and Himedia (India) respectively, all the reagents were used without further purification. Deionized water was employed as the solvent.

The calcium and phosphate solutions were prepared by dissolving 0.5 M calcium nitrate tetrahydrate with 1% gelatin and 0.3 M di ammonium hydrogen phosphate with 6% of sodium alginate in water. pH of these solutions was brought above 10 using ammonia solution at room temperature. The phosphate solution was added drop-wise into the calcium solution under vigorous stirring and aged for 24 hrs. Then the resulting colloidal mixture was centrifuged and the precipitate was dried at room temperature and named as HA/Alg/Gel (Hydroxyapatite/Alginate/Gelatin)

9.2.2 SORPTION EXPERIMENT

Pb^{2+} and Cd^{2+} sorption experiment was performed by batch equilibrium method and the experiment was carried out for the adsorbent dosage of 0.1 g in aqueous solution at 37°C. The adsorbent was added to 50 mL of 2 mMl^{-1} $Pb(NO_3)_2$ and $Cd(NO_3)_2$ solutions of pH 5 and aliquots were collected for predetermined time intervals 1, 4, 7, 24 and 29 hrs. Then the experiment was terminated and the adsorbent was filtered off, washed with deionized water and dried at room temperature for measuring the Ca, Pb^{2+} and Cd^{2+} concentrations. The amount of Pb^{2+} and Cd^{2+} removed by unit mass of adsorbent at a given time is calculated by mass balance equation

$$q = \frac{v}{m}(c_0 - c) \, mM/g \qquad (1)$$

where v is the volume of the solution in liters, m is the mass of the adsorbent in grams, c_0 and c are the initial and final lead and cadmium concentrations in the solution in millimols per liter.

9.2.3 CHARACTERIZATION

The phase composition and crystallographic structure of the adsorbents before and after sorption were identified by X-ray diffraction (PANalytical X' Pert PRO diffractometer) using CuKα radiation (1.5406 Å) with voltage and current setting of 40 kV and 30 mA respectively. Inductively coupled plasma optical emission spectroscope (ICP-OES, Perkin Elmer Optima 5300 DV) was used to determine the Ca, Pb^{2+} and Cd^{2+} concentrations.

9.2.4 KINETIC MODELS

The sorption kinetics of Pb^{2+} was analyzed on the basis of the two widely used kinetic models such as pseudo first and second order kinetic models.

9.2.4.1 The Pseudo-First Order Kinetic Model

The first order Lagrangian equation pertaining to the adsorption rate is based on the adsorption capacity and its linearized form can be expressed as

$$\log(q_e - q_t) = \log q_e - \frac{k_1}{2.303}t \tag{2}$$

where q_t is the Pb^{2+} adsorbed at time t in millimol per gram, q_e is the amount of Pb^{2+} adsorbed at equilibrium in millimol per gram and k_1 is the adsorption rate constant. The slope of the straight line plots of $\log(q_e - q_t)$ against t will give the value of q_e, rate constant and regression coefficient.

9.2.4.2 The Pseudo-Second Order Kinetic Model

The linearized second order kinetic model can be expressed as

$$\frac{t}{q_t} = \frac{1}{k_2 q_e^2} + \frac{1}{q_e}t \tag{3}$$

where q_t and q_e are the Pb^{2+} adsorbed on the surface of the adsorbents at time t and at equilibrium in millimols per gram respectively and k_2 is the rate constant. The slope of the straight-line plots of t/q_t against t will give the value of q_e and rate constant.

9.2.5 SORPTION ISOTHERM

Adsorption isotherm studies are essential for elucidating the adsorption process at equilibrium conditions. An adsorption isotherm is character-ized by certain constants, which express the affinity of the adsorbent and can also be used to find the adsorption capacity of the sorbent. Two most widely used mathematical models Langmuir and Freundlich adsorption isotherms were adopted for expressing the quantitative relationship between the extent of sorption and the residual solute concentration. Langmuir adsorption isotherm assumes the monolayer adsorption process on a homogeneous surface and the equation is expressed as

$$q = \frac{q_{max}bc}{1+bc} \qquad (4)$$

The linear form of Langmuir isotherm model can be expressed as

$$\frac{c}{q} = \frac{c}{q_{max}} + \frac{1}{q_{max}b}$$

where q and q_{max} represents the observed and maximum sorption capacities in millimol per gram and b is the Langmuir constant. The value of q_{max} and b can be determined from the slope and intercept of the linear plot of c/q versus c. The Freundlich equation is an empirical expression that encompasses the heterogeneity of the adsorbent surface and the exponential distribution of sites and their energies. It can be expressed as

$$q = kc^{\frac{1}{n}} \qquad (5)$$

The linear form of Freundlich equation can be expressed as

$$\ln q = \ln k + \frac{1}{n}\ln c$$

where k is a constant representing sorption capacity and n is a constant representing adsorption intensity parameter. The value of k and n can be determined from the slope and intercept of the linear plot of $\ln q$ versus $\ln c$.

9.3   RESULTS AND DISCUSSION

Figure 9.1 shows XRD pattern of adsorbent before sorption and the mineral phase present in the sample is confirmed to be HA (JCPDS card

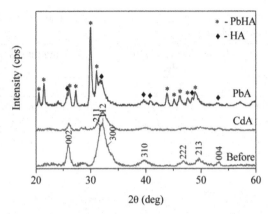

FIGURE 9.1   XRD pattern of the adsorbent before and after sorption.

No. 09–0432) and the crystallite size calculated by Debye-Scherrer approximation is 31 nm. After Pb^{2+} sorption, new peaks appeared along with the peaks of nHA and the additional phase present was identified to be hydroxypyromorphite (PbHA, $Pb_5(PO4)3OH$; card no. 24–0586). Cd^{2+} sorbed samples did not exhibit any new phase but it caused drastic reduction in the intensity of the characteristic HA peaks and disappearance of some peaks of HA.

Lagrangian pseudo first and second order rate equations were applied to experimental data. The rate constants k_2, Pb^{2+} adsorbed at equilibrium (q_e) and regression coefficient (R^2) obtained from the plots of pseudo second order rate equations and the values of Langmuir constants are given in Table 9.1. Based on the q_e and R^2 values, pseudo second order model was found to be more appropriate for both Pb^{2+} and Cd^{2+} sorption. It revealed that the process involved is chemisorption consisting a rapid phase in the initial 1 hr. followed by a linear phase in the next 6 hrs. and a slow removal phase until the 29 hrs. The amount of Pb^{2+} and Cd^{2+} removed (q) by the adsorbent deduced using equation (i) are depicted in Figs. 9.2 and 9.3.

The correlation regression coefficients calculated from Freundlich and Langmuir isotherm equations show that the adsorption process is better defined by Langmuir than by the Freundlich equation indicating the monolayer sorption (Fig. 9.4). The saturation capacity of HA/Alg/Gel are 0.7080 and 0.7621 mM/g for Pb^{2+} and Cd^{2+} respectively which confirmed that HA/Alg/Gel is a best candidate for the removal of Cd^{2+} than for Pb^{2+}.

TABLE 9.1 Experimental and Calculated Parameters of Pseudo Second Order Kinetic and Langmuir Isotherm Models

Metal ion	q_e (exp)	Second order kinetic model			Langmuir isotherm			
		q_e	k_2	R^2	q_{max}	b	R^2	k_L
Lead	0.9999	1.0219	1.5193	0.9999	0.7080	61.0910	0.9872	0.0081
Cadmium	0.9737	0.9916	1.6710	0.9996	0.7621	55.4313	0.9949	0.0089

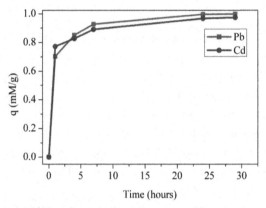

FIGURE 9.2 Pb^{2+} and Cd^{2+} removal capacity of the adsorbent with time.

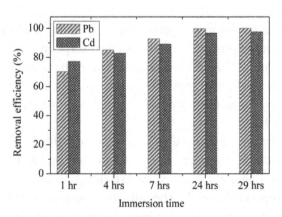

FIGURE 9.3 Removal efficiency of the adsorbent.

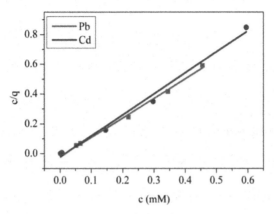

FIGURE 9.4 Linear plots of Langmuir isotherm.

Separation factor R_L is defined as $R_L=1/(1+bc_0)$. In this study R_L values calculated from Langmuir constant confirmed the favorable sorption ($0 < R_L < 1$) of lead and cadmium ions onto HA/Alg/Gel. It appears that a dissolution-precipitation mechanism dominates Pb^{2+} adsorption whereas a surface complexation mechanism may exist for Cd^{2+} sorption and this result was further confirmed by XRD.

9.4 CONCLUSIONS

nHA immobilized alginate-gelatin interpenetrating polymeric composite was prepared by wet chemical method and its adsorption capacity was studied for Pb^{2+} and Cd^{2+}. It was observed that equilibrium was attained in 7 hours for both the system. The sorption kinetics exhibited best fit to the pseudo-second order equation and the equilibrium followed Langmuir isotherm model. The mechanism of adsorption was found to be different for both Pb^{2+} and Cd^{2+}. The composite studied was found to be a potential adsorbent for both lead and cadmium.

ACKNOWLEDGMENT

The author K.S. acknowledges DST, India (Project Ref. No: SR/WOS-A/PS-15/2011) for financial support.

KEYWORDS

- alginate
- gelatin
- heavy metal
- hydroxyapatite

REFERENCES

1. Bailliez, S., Nzihou, A. The kinetics of surface area reduction during isothermal sintering of hydroxyapatite adsorbent. Chem Eng Jour., 98, 141–152 (2004).
2. Freundlich, H. M. F. Z. Phys. Chem., 57, 385–470 (1906).
3. Girija, E. K., Kumar, G. S., Thamizhavel, A., Yokogawa, Y., Kalkura, S. N. Role of material processing on the sinterability and thermal stability of nanocrystalline hydroxyapatite. Powder Technology. 225, 190–195 (2012).
4. Langmuir, I. The adsorption of gases on plane surfaces of glass, mica and platinum. J. Am. Chem. Soc. 40, 1361–1403 (1918).
5. Mavropoulos, E., Rossi, A. M., Costa, A. M., Perez, C. A. C., Moreira, J. C., Saldanha, M. Studies on the mechanisms of lead immobilization by hydroxyapatite. Environ.Sci. Technol. 36, 1625–1629 (2002).
6. Sangeetha, K., Thamizhavel, A., Girija, E. K. Effect of gelatin on the in situ formation of Alginate/Hydroxyapatite nanocomposites. Mater Lett. 91, 27–30 (2013).
7. Schröder, E., Jönsson, T., Poole, L. Hydroxyapatite chromatography: altering the phosphate-dependent elution profile of protein as a function of pH. Anal Biochem. 313, 176–178 (2003).
8. Stötzel, C., Müller, F. A., Reinert, F., Niederdraenk, F., Barralet, J. E., Gburecka, U. Ion adsorption behavior of hydroxyapatite with different crystallinities. Colloids Surf., B, 74, 91–95 (2009).
9. Sundaram, C. S., N. Viswanathan, N., Meenakshi, S. Defluoridation chemistry of synthetic hydroxyapatite at nano scale: equilibrium and kinetic studies. J. Hazard. Mater. 155, 206–215 (2008).
10. Tiselius, A., Hjertén, S., Levinö. Protein chromatography on calcium phosphate columns. Arch Biochem Biophys. 65, 132–155 (1956).
11. Xu, Y., Schwartz, F. W., Traina, S. J. Sorption of Zn^{2+} and Cd^{2+} on hydroxyapatite surfaces. Environ. Sci. Technol. 28, 1472–1480 (1994).

CHAPTER 10

MICROWAVE ASSISTED SYNTHESIS OF POLYACRYLAMIDE GRAFTED CASEIN (CAS-G-PAM) AS AN EFFECTIVE FLOCCULENT FOR WASTEWATER TREATMENT

SWETA SINHA, GAUTAM SEN, and SUMIT MISHRA

Department of Applied Chemistry, Birla Institute of Technology, Mesra, Ranchi – 835215, Jharkhand, India; Tel.: +91 9801334228; E-mail: sweta.sinha2203@gmail.com

CONTENTS

Abstract 154
10.1 Introduction 154
10.2 Materials and Methods 156
10.3 Results and Discussion 161
10.4 Conclusion 168
Acknowledgements 168
Keywords 169
References 169

ABSTRACT

Various grades of poly-(acryl amide) grafted casein (CAS-g-PAM) were synthesized by *microwave-assisted* method using chemical free radical initiator (e.g., ceric ammonium nitrate). The grafting of the poly (acryl amide) chains on the casein backbone was confirmed through intrinsic viscosity study and other physicochemical characterization techniques like FTIR spectroscopy, elemental analysis (C, H, N, O and S), scanning electron microscopy (SEM) and number average molecular weight determination through osmometry. The intrinsic viscosity and number average molecular weight of casein appreciably improved on grafting of poly (acryl amide) chains. Further, flocculation efficacy of the tailor-made graft copolymer was studied in kaolin suspension and through *jar test* procedures, towards possible application as flocculent. Further, flocculation efficacy of the 'best grade' (as determined by *'jar test'* in kaolin suspension) in wastewater was studied for possible application in reduction of pollutant load of wastewater.

10.1 INTRODUCTION

As the world's population is increasing, while the availability of potable water is decreasing. Since fresh water is essential for the survival of human beings and become a valuable commodity, recycling and reuse of wastewater becomes indispensable for all the processing industries to reduce the processing cost and also to abide by the environmental rules and regulations (Bolto and Gregory, 2007).

Wastewater contains solid particles with a wide variety of shapes, sizes, densities, etc. Specific properties of these particles affect their behavior in liquid phases and thus the removal capabilities. Man microbiological and chemical contaminants found in wastewater get adsorbed in these solid particles. Thus, an essential step for purification and recycling of wastewaters and industrial effluents is removal of these solid particles. Flocculation is a technique where polymers are involved in a solid–liquid separation by aggregation process of colloidal particles (Barkert and Hartmann, 1988; Chang et al., 2008). Flocculation aided by synthetic and semi synthetic flocculants (Pal et al., 2008; Sen et al., 2012) seems to be a highly effective strategy.

Casein makes up approximately 80 % of the protein in milk with the rest being whey proteins (Aschi, 2001). Besides its relevance as a nutritional product, casein has been used as the raw material for application in many fields such as man-made fibers, adhesives, glues, and paint binders (Purevsuren and Davaajav, 2001; Wang, et al., 2005). Though casein fiber is praised as the healthy and comfortable fiber, it still has some major drawbacks such as poor wet-rub resistance and susceptibility to microbial attack (Dong and Hsieh, 2000).

Nowadays, there is a growing interest in the attempts to combine the natural polymer with a synthetic polymer through graft copolymerization, since such treatments, in general, improve some of the disadvantages associated with these fibers with only negligible bulk changes of properties.

One of the most effective ways of modification of natural biopolymer (casein in this case) is by graft copolymerization with suitable monomers. Graft copolymerization on natural polymer has become an important resource for developing advanced materials as it can improve the functional properties of natural biopolymer (Da Silva et al., 2007; Geresh et al., 2004; Masuhiro et al., 2005; Singh et al., 2007). As a tailor-made material, grafted polymers find such diversified applications as matrices for controlled drug release (Nostrum et al., 2004; Singh et al., 2009; Rath and Singh, 2013), gums with improvised properties (Da Silva et al., 2007), and flocculants (Bharti et al., 2013; Sinha et al., 2013).

The properties of the end product (grafted casein) can be suitably modulated in terms of percentage grafting. The end product macromolecule is thus programmed at the molecular level for desired applications.

The most contemporary technique in graft copolymerization involves the use of microwave radiations to initiate the grafting reactions. Superiority of this technique over others has been well discussed in earlier studies (Mishra and Sen, 2011). Microwave radiations cause 'selective excitation' of only the polar bonds, leading to their rupture/cleavage – thus resulting in formation of free radical sites. The 'C-C' backbone of the preformed polymer being relatively non-polar, remains unaffected by the microwave radiation, thus the structural integrity of the backbone remains intact, leading to a superior product (Mishra and Sen, 2011). Microwave based graft copolymer synthesis is further classified into two types: *microwave initiated* synthesis (using microwave radiation alone to initiate grafting) and *microwave assisted* synthesis (using a synergism of microwave radiation and chemical free radical initiator to initiate grafting).

The most profound change due to grafting takes place in branching of the macromolecule and in drastic increase in hydrodynamic volume – both of which are the selection criteria for a good flocculent (Singh et al., 2000; Brostow et al., 2007).

The study described in this paper involve grafting of poly acrylamide (PAM) onto the backbone of casein, thus resulting in formation of poly acrylamide grafted casein (CAS-g-PAM). The synthesis has been carried out by *microwave-assisted* method, which involved a synergism of microwave radiations and ceric ammonium nitrate (CAN) to initiate the grafting reaction. The flocculation efficacies of the grafted product have been studied toward its application as flocculent for wastewater treatment.

10.2   MATERIALS AND METHODS

10.2.1   MATERIALS

Casein was procured from Loba Chemie Pvt Ltd, Mumbai, India. Acrylamide was procured from Sisco research laboratory, Mumbai, India. Ceric ammonium nitrate was supplied by E. Merck (India), Mumbai, India. Acetone was purchased from Rankem, New Delhi, India. All the chemicals were used as received; without further modification.

10.2.2   SYNTHESIS OF GRAFT COPOLYMER

Microwave assisted synthesis of poly acrylamide grafted casein (CAS-g-PAM), using CAN as free radical initiator: 1 g of casein was dissolved in 40 mL distilled water. Desired amount of acrylamide was dissolved in 10 mL of water and was added to the casein suspension, amid high speed stirring, followed by addition of catalytic amount of CAN. They were mixed under high-speed stirring (to affect a suspension of acrylamide in the reaction mixture) and were transferred to the reaction vessel (1000 mL borosil beaker) and transferred to the microwave reactor

(Catalyst™ system CATA 4 R). In the reactor, stirring was continued and irradiated with microwave radiation (700 Watts) until gelling sets in, keeping the irradiation cut-off temperature at 70 degree centigrade. Once the microwave irradiation procedure got complete, the gel like mass left in the reaction vessel was cooled and poured into excess of acetone. The precipitated grafted polymer was collected and was dried in hot air oven. Subsequently, it was pulverized and sieved. The synthesis details of various grades of the graft copolymer have been shown in Table 10.1. The percentage grafting of this microwave assisted synthesized CAS-g-PAM was evaluated as:

$$\% \text{ Grafting } = \frac{\text{Wt. of graft copolymer - Wt. of Protein}}{\text{Wt. of Protein}} \times 100$$

10.2.3 PURIFICATION OF THE GRAFT COPOLYMER BY SOLVENT EXTRACTION METHOD

Any occluded PAM formed due to competing homopolymer formation reaction was removed by extraction with acetone for 24 hours.

10.2.4 CHARACTERIZATION

10.2.4.1 Intrinsic Viscosity Measurement

Viscosity measurements of the polymer solutions were carried out with an Ubbelodhe viscometer (Constant: 0.003899) at 25°C. The viscosities were measured in aqueous solutions at neutral pH. The time of flow was measured at 0.1, 0.05, 0.025, and 0.0125 wt.%. The intrinsic viscosity was calculated by plotting η_{sp}/C versus C and ln η_{rel}/C versus C and then taking the common intercept at C = 0 of the best fitted straight lines from the two sets of points as described by Sen et al. (2009). The intrinsic viscosity thus evaluated for various grades of the graft copolymer has been reported in Table 10.1.

TABLE 10.1 Synthesis Details of Poly Acrylamide Grafted Casein (CAS-g-PAM)

Polymer Grade	Wt. of Casein (g)	Wt. of acrylamide (g)	Wt. of CAN (g)	Time of irradiation (up to gel formation in min)	percentage grafting (% G)	Number Average Molecular weight (kDa)	Intrinsic viscosity (dl/g)
CAS-g-PAM 1	1	10	0.1	>3		620	2.58
CAS-g-PAM 2	1	10	0.2	1.5		4485	9.39
CAS-g-PAM 3	1	10	0.3	2		1238	6.90
CAS-g-PAM 4	1	5	0.2	2.5		335	1.35
CAS-g-PAM 5	1	15	0.2	2		1770	7.28
Casein	-	-	-	-		30	0.89

10.2.4.2   Elemental Analysis

The elemental analysis of Casein and that of CAS-g-PAM 2 (best grade) was carried out by an Elemental Analyzer (Make – M/s Elementar, Germany; Model – Vario EL III). The estimation of three elements, that is, carbon, hydrogen and nitrogen were undertaken. The results have been summarized in Table 10.3.

10.2.4.3   Solubility in Aqueous Solution

Solubility determination for casein (starting material) and of the synthesized grafted grades of casein was determined by gravimetric method, at 25°C in distilled water.

10.2.4.4   FTIR Spectroscopy

The FTIR spectrums of Casein and of CAS-g-PAM 2 (Fig. 10.1) were recorded in solid state, by KBr pellet method, using FTIR spectrophotometer

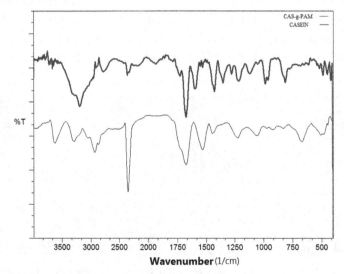

FIGURE 10.1   FTIR spectrum of Casein and CAS-g-PAM 2.

(Model IR-Prestige 21, Shimadzu Corporation, Japan) in the fingerprinting region of IR (400 and 4000 cm^{-1}).

10.2.4.5   Scanning Electron Microscopy

Surface morphology of Casein (Fig. 10.2(a)) and CAS-g-PAM 2 (Fig. 10.2(b)) were analyzed by scanning electron microscopy (SEM) in powdered form (Model: JSM-6390LV, Jeol, Japan).

10.2.4.6   Determination of Number Average Molecular Weight

The number average molecular weights of casein and various grades of CAS-g-PAMs were determined in aqueous medium by Osmometry (A+ Adv. Instruments, INC. Model 3320, Osmometer).

10.2.5   FLOCCULATION STUDIES

10.2.5.1   Flocculation Study in Kaolin Suspension

Flocculation efficacy of all the synthesized grades of CAS-g-PAM and that of casein were studied by standard 'jar test' procedure, in 0.25% kaolin suspension. All flocculation experiments were carried out in 'jar test' apparatus (Make: Simeco, Kolkata, India). The test protocol

FIGURE 10.2   (a) SEM morphology of Casein; (b) SEM morphology of CAS-g-PAM 2.

involved taking a measured quantity (800 mL) of the 0.25% kaolin suspension in 1L borosil beaker. Calculated amount of the flocculent (Casein or various grades of CAS-g-PAM) was added in concentrated solution form to achieve the desired dosage (ranging from 0 ppm to 1.25 ppm). The solutions were identically stirred (in 'jar test' apparatus), at 150 rpm for 30s, 60 rpm for 5 min, followed by 15 min of settling time. Afterwards, supernatant liquid was collected and turbidity measured in a calibrated nephelo-turbidity meter (Digital Nephelo-Turbidity Meter 132, Systronics, India). The flocculation efficacy thus studied for casein and various grades of CAS-g-PAM at various flocculent dosages have been reported.

10.2.5.2   Flocculation Study in Sewage Wastewater

Flocculation efficacy of the best grade of the grafted casein (CAS-g-PAM 2) in municipal wastewater, at the optimized dosage as determined by earlier flocculation experiment (by 'jar test' procedure as in Section 10.3.5.1, but with wastewater instead of kaolin suspension), was studied and compared to the efficacy in case of the starting material (casein). The flocculation efficacy was evaluated by assessment of pollutant load in the collected supernatant. The experiment was done in three sets.

SET 1: Wastewater without flocculent; SET 2: Wastewater with 0.75 ppm of Casein; SET 3: Wastewater with 0.50 ppm of CAS-g-PAM 2.

The water quality of these supernatants was analyzed by standard procedures (Greenberg, 1999), as reported in Table 10.2.

10.3   RESULTS AND DISCUSSION

10.3.1   SYNTHESIS OF CAS-G-PAM BY MICROWAVE ASSISTED METHOD

Polyacrylamide grafted casein has been synthesized by microwave assisted method. The term *microwave assisted* has been coined by earlier studies (Odian, 2002; Mishra et al., 2012; Pal et al., 2012). It refers to

TABLE 10.2 Comparative Study of Performance Best Grades of CAS-g-PAM for the Treatment of Municipal Sewage Wastewater

Parameter	SET 1 (i.e., wastewater without flocculent) Supernatant liquid	SET 2 (i.e., wastewater with 0.75 ppm of Casein) Supernatant liquid	SET 3 (i.e., wastewater with 0.50 ppm of CAS-g-PAM 2) Supernatant liquid
Total solid (ppm)	2040	1870	1440
TDS (ppm)	1810	1660	1280
TSS (ppm)	230	210	160
COD (ppm)	322	290	114
Ca^{2+} (ppm)	753	647	500.30
Cr(IV) (ppm)	2.98	1.601	0.447

a process of graft copolymer synthesis, which is a hybrid of microwave initiated and conventional method (chemical free radical initiator based method) of synthesis, that is, it is based on free radical mechanism using microwave radiation in synergism with chemical free radical initiator (ceric ammonium nitrate) to generate free radical sites on the protein backbone. Various grades of the graft copolymer were synthesized by varying the ceric ammonium nitrate (CAN) and acrylamide (monomer) concentration. In each case, the microwave irradiation of the reaction mixture was continued until it sets into a viscous gel like mass (or up to 3 min if no gelling took place), the microwave irradiation cutoff temperature being maintained at 70 degree centigrade throughout the process. The synthesis details have been tabulated in Table 10.1. The optimized grade has been determined through its higher percentage grafting, intrinsic viscosity and molecular weight. The approach of synthesis involved optimization with respect to CAN, keeping the acrylamide concentration constant (i.e., CAS-g-PAM 1, 2, 3); followed by optimization with respect to acrylamide, keeping the CAN concentration as optimized before (i.e., CAS-g-PAM 2, 4, 5). From Table 10.1, it is obvious that the grafting is optimized at acrylamide concentration of 10 g and CAN concentration of 0.2 g in the reaction mixture, when the microwave power is maintained at 700 W.

Ceric ammonium nitrate is electron deficient molecule, so it takes electrons from alcoholic oxygen in casein to form a new bond, that is, Ce-O. This bond being more polar (than O-H bond) breaks easily in the presence of microwave irradiation to form free radical site on the backbone of casein, from where the graft chains grow. The proposed mechanism of microwave-assisted synthesis has been hypothesized in detail in earlier studies (Mishra et al., 2011; Pal et al., 2011). The proposed mechanism of microwave assisted grafting has been depicted in Scheme 10.1.

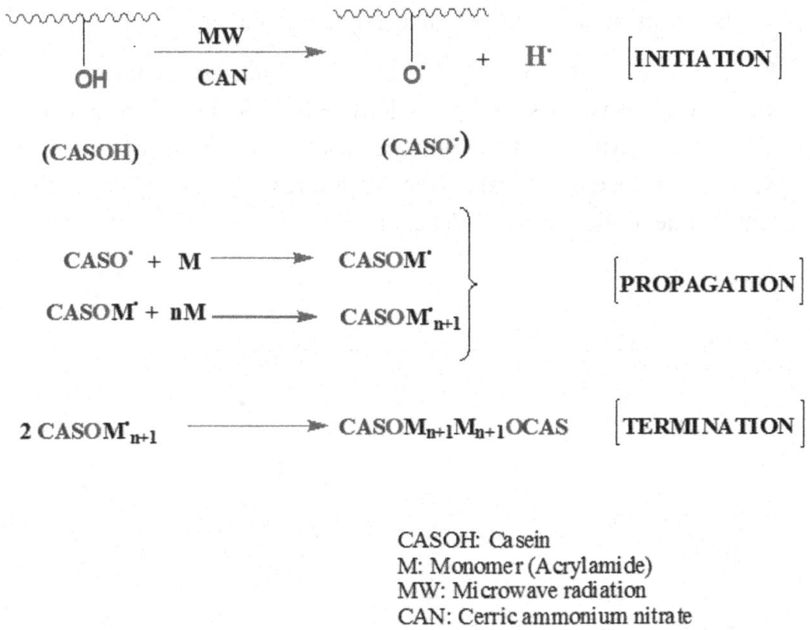

SCHEME 10.1 Schematic representation of mechanism for synthesis of CAS-g-PMMA using 'microwave assisted method'.

10.3.2 CHARACTERIZATION

10.3.2.1 Estimation and Interpretation of Intrinsic Viscosity

The intrinsic viscosity was evaluated for casein and the various grades of

CAS-g-PAM, as shown in Table 10.1. Intrinsic viscosity is practically the hydrodynamic volume of the macromolecule in solution (aqueous solution in this case). It is obvious from Table 10.1 that the intrinsic viscosities of all the grades of CAS-g-PAM are greater than that of casein. This can be explained by the increase in hydrodynamic volume due to the grafting of the PAM chains on the main polymer backbone, that is, casein. The grafted PAM chains increase hydrodynamic volume by two ways:

1. By uncoiling of the protein chain through steric hindrance to intra-molecular bonding.
2. By contributing their own hydrodynamic volume.

Further, this is in good agreement with Mark–Houwink–Sakurada relationship (intrinsic viscosity $\eta = KM^\alpha$ where K and α are constants, both related to stiffness of the polymer chains), applying which we can explain the increase in intrinsic viscosity as a result of increase in molecular weight due to the grafted PAM chains.

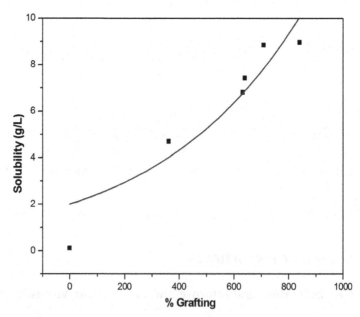

FIGURE 10.3　Solubility (weight in g/L vs. % grafting) plot of casein and different grades of CAS-g-PAM in aqueous solution.

As evident from Fig. 10.3, intrinsic viscosity has positive correlation with percentage grafting.

10.3.2.2   Elemental Analysis

The results of elemental analysis for casein and that of the best grade of polyacrylamide-grafted casein (i.e., CAS-g-PAM 2) are given in Table 10.3. Casein, which is a starting material, has low percentages of nitrogen (i.e., 12.77%) than CAS-g-PAM 2. The higher percentage of nitrogen is contributed by the grafted PAM chains (which have higher nitrogen content than the casein).

10.3.2.3   Solubility in Aqueous Solution

The solubility details of Casein and different grades of CAS-g-PAM in aqueous solution (polar solvent) have been shown in Fig. 10.3. Since casein is an amphiphilic protein shows partial solubility in both types of solvents, that is, polar as well as in non polar. Grafting of polar monomers (i.e., acrylamide) onto the backbone of casein has been found to give rise to improved solubility (Fig. 10.3) by virtue of the incorporated polar groups. All the grades of grafted casein have shown better solubility than the starting material (casein) in aqueous solution. Consequently, higher the percentage grafting of the casein, higher is its solubility in aqueous solution, that is, the best grade (the grade with highest percentage grafting) have highest solubility.

TABLE 10.3   Elemental Analysis

Polymer grade	% C	% H	% N
Casein	47.66	7.778	12.77
PAM	50.08	7.69	19.76
CAS-g-PAM 3	48.70	7.68	17.82

10.3.2.4   FTIR Spectroscopy

Figure 10.1 shows the FT-IR spectra of starting material casein and best grade of polyacrylamide-grafted casein (CAS-g-PAM 2), respectively. Major regions of the FTIR spectra of casein showed absorption peaks at 3626.17 cm^{-1} (O-H stretching vibrations), 2924.09 cm^{-1} (C-H stretching vibrations), 1674.21 cm^{-1} (C=O stretching), 1531.48 cm^{-1} (N-H bending), 1442.75 cm^{-1} (aromatic C=C stretching), and 1230.58 cm^{-1} (saturated C-C stretching).

In case of CAS-g-PAM 2, the peak at 3188 cm^{-1} is due to the overlapping of O–H stretching band of hydroxyl group of casein and N–H stretching band of amide group of PAM with each other. The additional peak of CAS-g-PAM 2 at 963 cm^{-1} is assigned to C-O-C asymmetric stretching vibrations. C-O-C bond is present in grafted casein and is absent in both casein as well as PAM and thus is the confirmation of grafting.

10.3.2.5   Scanning Electron Microscope

It is evident from the SEM micrographs of casein (Fig. 10.2(a)) and that of the best grade of CAS-g-PAM (Fig. 10.2(b)) that profound morphological change have taken place. The homogeneous morphology of CAS has been transformed to heterogeneous (fibrillar) morphology

10.3.2.6   Number Average Molecular Weight of the Polymers

As evident from Table 10.1, higher the percentage grafting, higher is the number average molecular weight. Thus, there is predictable correlation between the three parameters, that is, percentage grafting (%G), intrinsic viscosity (η), & number average molecular weight. The correlations have been depicted in Fig. 10.4.

10.3.3   FLOCCULATION STUDY

The flocculation efficacy of the best grade of grafted Casein (CAS-g-PAM 2) and that of Casein were studied in municipal wastewater at dosage optimized as in kaolin suspension study, by standard 'Jar test' procedure (Fig.

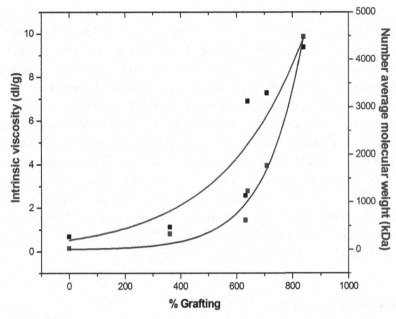

FIGURE 10.4 Co-relation graph of intrinsic viscosity, percentage grafting and number average molecular weight of casein and different grades of CAS-g-PAM.

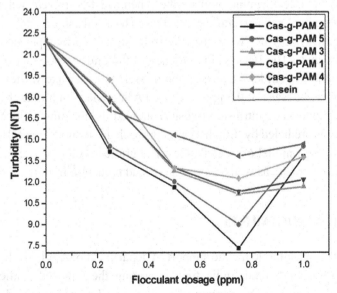

FIGURE 10.5 Flocculation profile, Jar test (in 0.25% kaolin suspension).

10.5). CAS-g-PAM 2 was found to be able to considerably reduce the pollutant load of wastewater, compared to Casein, as evidenced by analysis of supernatants drawn from the 'jar test' procedure at optimized flocculent dosage (0.5 ppm). A comparative study (Table 10.3) of water quality of supernatants drawn from 'jar test' procedure in case of wastewater alone (SET 1), wastewater with 0.75 ppm (optimized dose) of Casein as flocculent (SET 2) and wastewater with 0.5 ppm of CAS-g-PAM 2 as flocculent (SET 3) have shown much better water quality in case of SET 3. Drastic reduction in metal content (Chromium VI & Total iron) and appreciable reduction in organic load (in terms of COD) were observed. Thus, CAS-g-PAM is much better flocculent than the starting material (Casein). The fact that reduction in total suspended solid (TSS) were much higher compared to the reduction in total dissolved solid (TDS) indicated the underlying mechanism to be flocculation and not adsorption.

10.4 CONCLUSION

Polyacrylamide grafted casein (CAS-g-PAM) has been synthesized by microwave assisted technique, which involved a synergism of microwave radiation and ceric ammonium nitrate (chemical free radical initiator) to initiate the free radical grafting reaction. The synthesized grades of this novel graft copolymer were characterized through various physicochemical techniques. The flocculation efficacy of the graft co polymer has been appreciable in both kaolin suspension as well as in sewage wastewater, at an optimized dosage of 0.5 ppm. CAS-g-PAM grade with highest hydrodynamic volume (i.e., intrinsic viscosity) showed the maximum flocculation efficacy, as predicted by 'Singh's easy approachability model' and 'Brostow, Pal and Singh model of flocculation.' The grafted protein was effective in TDS, heavy metal, as well as for organic load reduction from wastewater.

ACKNOWLEDGEMENTS

The authors acknowledge the support of department of Applied Chemistry BIT Mesra, Jharkhand, India for performing the reported studies. The financial support of Department of Science and Technology (DST) Govt.

of India, in form of DST fast tract grant (vide grant no. GOI SERB PROJ 08/12–09141) is also being acknowledged. Finally, the role of staff of Central Instrumentation Facility (CIF 1) of BIT Mesra is also being acknowledged.

KEYWORDS

- **casein**
- **flocculent**
- **microwave assisted synthesis**
- **tailor-made graft co polymer**

REFERENCES

1. Aschi, A., Gharbi, A., Bitri, L., Calmettes, P., Daoud, M., Beghin, V. A., Douillard, R. Langmuir, 17, 1896 (2001).
2. Barkert, H.; Hartmann, J. In Encyclopedia of Industrial Chemistry; John Wiley & Sons: New York, 1988; Vol. 5, p 251.
3. Bharti, S., Mishra, S., Sen, G. (2013). Ceric ion initiated synthesis of polyacrylamide grafted Oatmeal: Its applications as flocculent for waste water treatment, Carbohydrate polymer, 93, 528–536.
4. Bolto, B., Gregory, J. (2007). Organic polyelectrolytes in water treatment. Water Research, 41, 2301–2324.
5. Brostow, W., Pal, S., Singh, R. P. (2007). A model of flocculation. Materials Letters, 61, 4381–4384.
6. Brostow, W., Pal, S., Singh, R. P. (2007). A model of flocculation. Materials Letters, 61, 4381–4384.
7. Chang, Q.; Hao, X.; Duan, L. J Hazard Mater 2008, 159, 548.
8. Collins, E. A., Bares, J., Billmeyer, F. W. (1973). Experiments in polymer science (pp. 394–399). New York: John Wiley & Sons.
9. Da Silva, D.A., de Paula, R.C.M., Feitosa, J.P.A. (2007). Graft copolymerization of acrylamide onto cashew gum. Eur. Polym. J. 43, 2620–2629.
10. Dong Q. Z., Hsieh, Y. L. J. Appl. Polym. Sci., 77, 2543 (2000).
11. Geresh, S., Gdalevsky, G.Y., Gilboa, I., Voorspoels, J., Remon, J.P., Kost, J. (2004). Bioadhesive grafted starch copolymers as platforms for peroral drug delivery: a study of theophylline release. J. Controlled Release. 94, 391–398.
12. Gowariker, V. R., Viswanathan, N. V., Sreedhar, J. (1986). Polymer science. New Age International (p) Ltd.

13. Izydorczyk, Biliaderis, Cereal arabinoxylans: advances in structure and physico-chemical properties, Carbohydr. Polym, 28 (1995) 33–48.

14. Kaith, K. Kumar, In air synthesis of Psy-cl-poly(AAm) network and its application in water-absorption from oil-water emulsions, express Polym. Lett. 1 (2007) 474–480.

15. Kongparakul, S., Prasassarakich, P., Rempel, L. G. (2008). Effect of grafted methyl methacrylate on the catalytic hydrogenation of natural rubber. European Polymer Journal, 44, 1915–1920.

16. Masuhiro, T., Shafiul, I., Takayuki, A., Alessandra, B., Giuliano, F. (2005). Micro-wave irradiation technique to enhance protein fiber properties. AUTEX Res. J. 5, 40–48.

17. Mishra, A., Rajani, S., Agarwal, M., Dubey, R. (2002). P.psyllium-g-polyacrylamide: Synthesis and characterization. Polymer Bulletin, 48, 439–444.

18. Mishra, A., Srinivasan, R., Gupta, R. (2003). P. psyllium-g-polyacrylonitrile: synthe-sis and characterization. Colloid Polym Science, 281, 187–189.

19. Mishra, S., Sen, G. (2011). Microwave initiated synthesis of polymethylmethacrylate grafted guar (GG-g-PMMA), characterizations and applications. International Jour-nal of Biological Macromolecules, 48, 688–694.

20. Mishra, S., Mukul, A., Sen, G., Jha, U. (2011). Microwave assisted synthesis of poly-acrylamide grafted starch (St-g-PAM) and its applicability as flocculent for water treatment. International Journal of Biological Macromolecule, 48, 106–111.

21. Mishra, S., Rani, U., Sen, G. (2012). Microwave initiated synthesis and application of polyacrylic acid grafted carboxymethyl cellulose. Carbohydrate Polymers, 87, 2255–2262.

22. Mishra, S., Sen, G., Rani, U., Sinha, S. (2011). Microwave assisted synthesis of poly-acrylamide grafted agar (Ag-g-PAM) and its application as flocculent for wastewater treatment. International Journal of Biological Macromolecules, 49, 591- 598.

23. Nostrum, C.F.V., Veldhuis, T.F.J., Bos, G.W., Wim, E., Hennink, W.E. (2004). Tuning the degradation rate of poly (2-hydroxypropyl methacrylamide)-graft-oligo (lactic acid) stereocomplex hydrogels. Macromolecules 37, 2113–2118.

24. Odian, G. (2002). Principles of polymerization (3rd ed.). New York: John Wiley & sons.

25. Pal, S., Ghorai, S., Dash, M. K., Ghosh, S., Udayabhanu, G. (2011). Flocculation properties of polyacrylamide grafted carboxymethyl guargum (CMG-g-PAM) syn-thesized by conventional and microwave assisted method. Journal of Hazardous Ma-terial, 192, 1580–1588.

26. Pal, S., Sen, G., Ghosh, S., Singh, R. P. (2012), High performance polymeric floc-culants based on modified polysaccharides—Microwave assisted synthesis. Carbo-hydrate Polymers, 87, 336–342.

27. Pal, S., Sen, G., Karmakar, N. C., Mal, D., Singh, R. P. (2008). High performance flocculating agents based on cationic polysaccharides in relation to coal fine suspen-sion. Carbohydrate Polymers, 74, 590–596.

28. Petchetti L., Frishman W. H., Petrillo R., Raju K. (2007), Nutriceuticals in cardiovas-cular disease: psyllium, Cardiol Rev, 15,116–22.

29. Purevsuren B., Davaajav, Y. Journal of Thermal Analysis and Calorimetry, 65, 147 (2001).

30. Rakel. Rakel. Integrative Medicine, 2nd ed. Philadelphia, PA: Saunders Elsevier. 2007, 43.

31. Rani, P., Sen, G., Mishra, S., Jha, U. (2012). Microwave assisted synthesis of polyacrylamide grafted gum ghatti and its application as flocculent. Carbohydrate Polymers, 89, 275–281.

32. Rath, S.K., Singh, R.P. (2000). Flocculation characteristics of grafted and ungrafted starch amylase and amylopectin. J. Appl. Polym. Sci. 66, 1721–1729

33. Ruehrwein, R. A., Ward, D. W. (1952). Mechanism of clay aggregation by polyelectrolytes. Soil Science, 73, 485–492.

34. Saper RB, Eisenberg DM, Phillips RS. Common dietary supplements for weight loss. Am Fam Physician. 2004 Nov 1;70(9):1731–8. Review.

35. Sartore G, Reitano R, Barison A, Magnanini P, Cosma C, Burlina S, Manzato E, Fedele D, Lapolla A. The effects of psyllium on lipoproteins in type II diabetic patients. Eur J Clin Nutr. 2009;63(10):1269–71.

36. Sen, G., Pal, S. (2009a). Polyacrylamide grafted carboxymethyl tamarind (CMT-g-PAM): Development and application of a novel polymeric flocculent. Macromolecular Symposium, 277, 100–111.

37. Sen, G., Ghosh, S., Jha, U., Pal, S. (2011). Hydrolyzed polyacrylamide grafted carboxymethylstarch (Hyd CMS-g-PAM): An efficient flocculant for the treatment of textile industry wastewater. Chemical Engineering Journal, 171, 495–501.

38. Sen, G., Kumar, R., Ghosh, S., Pal, S. (2009). A novel polymeric flocculant based on polyacrylamide grafted carboxymethylstarch. Carbohydrate Polymers, 77, 822–831.

39. Sen, G., Mishra, S., Jha, U., Pal, S. (2010). Microwave initiated synthesis of polyacrylamide grafted guargum (GG-g-PAM)—Characterizations and application as matrix for controlled release of 5-amino salicylic acid. International Journal of Biological Macromolecules, 47, 164–170.

40. Sen, G., Mishra, S., Usha Rani, G., Rani, P., Prasad, R. (2012). Microwave initiated synthesis of polyacrylamide-grafted psyllium (Psy-g-PAM) and its application as flocculent. International Journal of Biological Macromolecules, , 50, 369–375.

41. Sen, G., Mishra, S., Usha Rani, G., Rani, P., Prasad, R. (2012). Microwave initiated synthesis of polyacrylamide grafted psyllium (Psy-g-PAM) and its application as flocculent. International Journal of Biological Macromolecules, 50, 369–375.

42. Sen, G., Pal, S. (2009b). Microwave initiated synthesis of polyacrylamide-grafted carboxymethylstarch (CMS-g-PAM): Application as a novel matrix for sustained drug release. International Journal of Biological Macromolecules, 45, 48–55.

43. Sen, G., Singh, R. P., Pal, S. (2010). Microwave-initiated synthesis of polyacrylamide grafted sodium alginate: Synthesis and characterization. Journal of Applied Polymer Science, 115, 63–71.

44. Singh, R. P. (1995). Advanced drag reducing and flocculating materials based on polysaccharides. In N. Prasad, J. E. Mark, & T. J. Fai (Eds.), Polymers and other advanced materials: Emerging technologies and business opportunities (pp. 227–249). New York: Plenum Press.

45. Singh, R. P., Karmakar, G. P., Rath, S. K., Karmakar, N. C., Pandey, S. R., Tripathy, T., et al. (2000). Biodegradable drag reducing agents and flocculants based on polysaccharides: Materials and applications. Polymer Engineering and Science, 40, 46–60.

46. Singh, R. P., Karmakar, G. P., Rath, S. K., Karmakar, N. C., Pandey, S. R., Tripathy, T., et al. (2000). Biodegradable drag reducing agents and flocculants based on polysaccharides: Materials and applications. Polymer Engineering & Science, 40, 46–60.

47. Singh, R. P., Pal, S., Krishnamoorthy, S., Adhikary, P., Ali, S.K. (2009). High technology and materials based on modified polysaccharides. Pure Appl. Chem. 81, 525–547.

48. Singh, V., Tiwari, A., Pandey, S., Singh, S.K. (2007). Peroxydisulfate initiated synthesis of potato starch-graft-poly (acrylonitrile) under microwave irradiation; eX-PRESS. Polym. Lett. 1, 51–58.

49. Sinha, S., Mishra, S., Sen, G. (2013). Microwave initiated synthesis of polyacrylamide grafted Casein (CAS-g-PAM) – Its application as a flocculent. International Journal of Biological Macromolecules, 60, 141–147.

50. Sinha, S., Mishra, S., Sen, G. (2013). Microwave initiated synthesis of polyacrylamide grafted Casein (CAS-g-PAM) – Its application as a flocculent. International Journal of Biological Macromolecules, 60, 141–147.

51. Usha Rani, G., Mishra, S., Sen, G., Jha, U. (2012). Polyacrylamide grafted agar: Synthesis and applications of conventional and microwave assisted technique. Carbohydrate Polymers, 90, 784–791.

52. Wang, N. G., Zhang, L. N., Lu, Y. S., Du, Y. M. J. Appl. Polym. Sci., 91, 332 (2005).

CHAPTER 11

POLYMER ASSISTED SYNTHESIS OF CDS NANOSTRUCTURE FOR PHOTOELECTROCHEMICAL SOLAR CELL APPLICATIONS

S. A. VANALAKAR,[1] J. H. KIM,[1] and P. S. PATIL[2]

[1]Department of Materials Science and Engineering, Chonnam National University, Gwangju–500 757, South Korea

[2]Thin Film Materials Laboratory, Department of Physics, Shivaji University, Kolhapur-416204, M.S., India

CONTENTS

Abstract 174
11.1 Introduction 174
11.2 Materials and Methods 181
11.3 Result and Discussion 183
11.4 Conclusion 197
Acknowledgment 198
Keywords 198
References 199

ABSTRACT

This chapter provides an overview of recent advances in the polymer-assisted synthesis of cadmium sulfide (CdS) nanostructures. Among wide band gap semiconductors CdS with its direct band gap of 2.42 eV at room temperature is a promising material in solar cell, photo-catalysis, chemical sensing, thermo-electricity, etc. The thin films of CdS are conventionally grown by physical or chemical deposition route. However, physical deposition route requires the high cost of the necessary equipment and restrictions of coatings on a relatively small area have limited their potential applications. Chemical-solution depositions such as sol–gel are more cost-effective, but many metal chalcogenides cannot be deposited and the control of stoichiometry is not always possible owing to differences in chemical reactivity among the metals. Here we report a novel process to grow metal chalcogenides films in large areas at low cost using polymer-assisted chemical bath deposition, where the polymer controls the viscosity and binds metal ions, resulting in a homogeneous distribution of metal precursors in the solution and the formation of uniform metal–organic films. The latter feature makes it possible to grow simple and complex crack-free epitaxial metal-chalcogenide thin films like CdS.

11.1 INTRODUCTION

Nanomaterials are keystones of nano-science and technology. Nanostructure science and technology is an extensive and interdisciplinary area of research and development activity that has been growing explosively worldwide in the past few years. It has the potential for revolutionizing the ways in which materials and products are created and the range and nature of functionalities that can be accessed. It is already having a significant commercial impact, which will undoubtedly increase in the future. Nanoscale materials are defined as a set of substances where at least one dimension is less than approximately 100 nanometers. A nanometer is one millionth of a millimeter – approximately 100,000 times smaller than the diameter of a human hair. Nanomaterials are of interest because at this scale unique optical, magnetic, electrical, and other properties arise. These emergent properties have the potential for great impacts in electronics, mechanical, environment, defense, medicine, and other fields. Some nanomaterials occur naturally, but of particular interest are engineered nanomaterials, which are designed for, and already being used in many commercial products and processes. Engineered nanomaterials are resources designed at the molecular (nanometer) level to take advantage of their small size and novel properties which are generally not seen in their conventional, bulk counterparts. The two main reasons why materials at the nano scale can have different properties are increased relative surface area and quantum effects. Nanomaterials have a much greater surface area to volume ratio than their conventional forms, which can lead to greater chemical reactivity and affect their strength. Also at the nano scale, quantum effects can become much more important in determining the materials properties and characteristics, leading to novel optical, electrical and magnetic behaviors. In recent years, the syntheses of nanomaterials have attracted considerable attention due to their unique properties. Following graph (Fig. 11.1) shows comparative study of publications on nanostructure from 1990 to 2013.

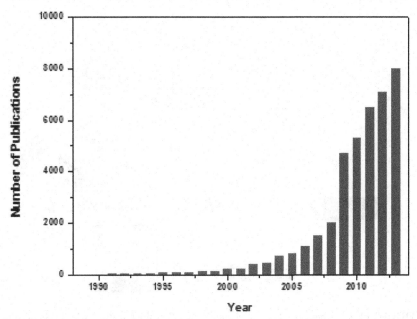

FIGURE 11.1   Number of publications per year (Word: Nanostructure; Source: web of knowledge).

A variety of inorganic nonstructural material can be synthesized from physical and chemical processing routes. The chemical processing route is particularly attractive due to its simplicity and versatility. This route includes chemical bath deposition, sol-gel, electrodeposition, mechano-chemical, screen printing, simple precipitation process, liquid phase deposition, successive ionic layer adsorption and reaction (SILAR) methods. Today, there has been a prevailing need for efficient, low temperature and low-cost deposition methods of material preparations for thin films as technological industrial applications. Chemical bath deposition (CBD) is a soft solution process that is capable of producing high-quality thin film at relatively low temperature. The fundamental CBD growth mechanism is similar to that of chemical vapor deposition (CVD), involving mass transport of reactants, adsorption, surface diffusion, reaction, nucleation and growth. Chemical bath deposition provides a simple and low-cost method to produce uniform, adherent and reproducible large-area thin films for thin-film electronics applications such as solar cells (Vanalakar et al., 2011). The schematic set up of CBD is given in Fig. 11.2. Historically, the first application of CBD was reported by Reynolds for the fabrication

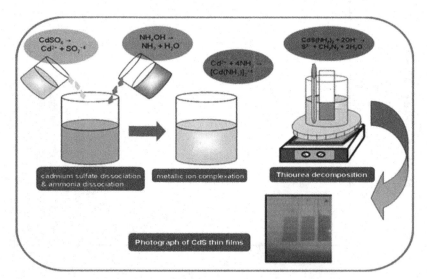

FIGURE 11.2 Schematic set up of chemical bath deposition method with CdS thin film preparation mechanism.

of lead sulfide photoconductive detectors in 1884. The reaction proceeded in a basic solution and a strongly adherent layer was formed on the interior surfaces of the reaction vessel. The first general review for CBD was reported by the Chopra et al. (1982). Several years later, a review by Lokhande (1991) was published with an emphasis on the deposition of metal chalcogenides. A comprehensive review was reported by Lincot et al. (1998), with a detailed analysis of the growth kinetics. Four more reviews by Nair et al. (1998), Savadogo (1998), Mane et al. (2000) and Hodes (2003) reported on their extensive work in the field with an emphasis on solar energy-related issues. Regardless of the deposition technique, the characterization of the post-deposited films and optimization of the deposition processes are still open subjects for research. A large number of studies have been carried out to achieve this goal in order to produce inorganic nanostructure thin films with good optoelectronic properties suitable for photovoltaic application. In this method, inorganic salts, polymer, pH, catalysis, surfactant and micelles acting as effective shape-controlling agents for nanostructure (Mai et.ai., 2010, Kariper et al., 2011).

Generally, inorganic nanostructure consists of a combination of elements of the II and VI groups of the periodic table of element (such as CdS, ZnS, CdSe, etc.). Generally, such inorganic combinations are referred as metal chalcogenides. Metal chalcogenides became an established versatile platform for multiple functional applications including solar cell, photocatalysis, chemical sensing, thermo-electricity, etc. (Moss et al., 1974, Lai et al., 2012, Tang et al., 2013, Gao et al., 2013). Among the metal chalcogenide compounds belonging to the II–VI family, cadmium sulfide (CdS) has received much attention due to the fact that it has a wide band gap of 2.42 eV at room temperature and it exhibits excellent properties for various optoelectronic applications within the visible range of the solar spectrum (Vanalakar et al., 2010, Roy et al., 2006). Cadmium sulfide has, like zinc sulfide, two crystal forms; the more stable hexagonal wurtzite structure (found in the mineral Greenockite) and the cubic zinc blende structure (found in the mineral Hawleyite) shown in Fig. 11.3. In both of these forms the cadmium and sulfur atoms are four coordinate (Wiberg et al., 2001, Abdolahzadeh et al., 2012). The chemical bonding of both phases is commonly described by sp^3 hybridization. In both structures, the coordination shell of the anion is made up of four cations in tetrahedral symmetry

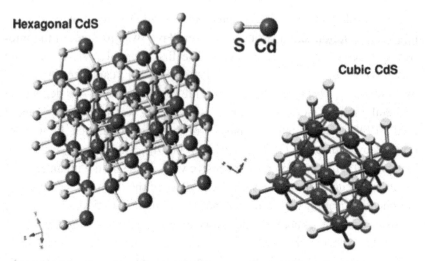

FIGURE 11.3   Crystal structures of CdS (hexagonal and cubic).

and the second neighboring shell is made up of 12 anions. One can look on these structures as having closely packed ions of S with Cd ions in tetrahedral voids. CdS have numerous applications in electro-optic devices such as laser materials, transducers, photoconducting cells, absorber layer, photosensors, optical wave-guides and non-linear integrated optical devices (Saxena et al., 2014) electrochemical cells, gas sensor (Trindade et al., 2001). Recently, CdS is also applied to dye-sensitized photoelectrochemical cells to improve their performance (Xiao et al., 2008, Vanalakar et al., 2011).

The use of polymers is a prominent method for the synthesis of CdS nanostructure. The reason is that the polymer matrices offer advantages like easy processability, solubility and control of the growth and morphology of the nanoparticles. Polymers have been used to facilitate the self-assembly process in forming various nanostructures and controlling the size and morphology of chalcogenide materials (Hanemann et al. 2010, Rezaul et al., 2010, Shang et al., 2009, Liang et al., 2008). The polymer controls the viscosity and binds metal ions, resulting in a homogeneous distribution of metal chalcogenide precursors in the solution and the formation of uniform thin films. The nature of the metal chalcogenide deposition is dominated by bottom-up growth, leading to ready formation of crack-free epitaxial metal chalcogenide and the ability to coat nanofeatured substrates in a conformal fashion (Burrell et al. 2008). As the chemical and physical properties of CdS depend on its composition, structure, phase, shape, size, and size distribution, so, morphology controlled synthesis and well-defined shapes for the formation of CdS are technologically motivating. Owing to high efficiency, simplicity and flexibility of their synthesis, bottom-up approaches using polymers, surfactants or micelles as the modifiable morphological/structural agents or templates are eagerly employed in the fabrication of one/two dimensional nanostructure (Zuo et al., 2012; Yu wt.al., 1999; Dhage et al., 2007; Yang et al., 2005). It is well known that the chemistry of surface modification of nanocrystals through the attachment of surface active agents may lead to the organized assembly of nanostructure (Bhirud et al., 2011). First report on improvement of polymer properties by incorporation of nanofillers was published in 1987. (Okada A. et al., 1987). Akbar et al. reported that the organized assembly could be realized in the case of ZnO by using four surfactant-Polyethylenimine (PEI), sodium dodecyl sulfate (SDS), polyethylene glycol (PEG), ethylene glycol (EG) and polyvinyl pyrrolidone (PVP) which probably bind selectively to the respective surface planes (Akbar et al., 2008). CdS nanoparticles embedded in different matrixes like polystyrene, polyvinyl alcohol, polyethylenimine, polypyrrole, polyethylene glycol etc. have been prepared by many workers (Xie et al., 1999; Sajinvi et al., 2000; Chen et al., 2000; Yu et al., 2001; Senthil et al., 2001; Lejmi et al., 2001; Du et al., 2002; Sone et al., 2002; Kippeny et al., 2002; Ma et al., 2003; Wu et al., 2004; Yao et al., 2005; Antolini et al., 2005; Wang et al., 2007; Gai et al., 2008; Khallaf et.ai., 2008; Li et al., 2009; Thongtem et al., 2009; Thambidurai et al., 2009; Li et al., 2009; Xiong et al., 2010; Cruz et al., 2010; Zhang et al., 2010; Jing et al., 2010; Zhang et al., 2012; Song et al., 2012; Song et al., 2014). In this chapter, we will summarize the CdS nanostructured prepared by polymer assistance as shown in Table 11.1.

Typically polymer and inorganic filler do not like each other – they are incompatible. So, the main task is how to concentrate them friendly, that is, how to distribute the nanoparticles in the matrix. Typically we would like to have a homogeneous distribution of nanoparticles. A big effort is made in the field of polymer nanocomposites to optimize and control the dispersion of nanoparticles – processing is very significant in that respect.

TABLE 11.1   Polymer Assistant CdS Nanostructures

Polymer	Nanostructure	Reference
Polypyrrole	Nanoparticles	Zhang et al. (2012)
Polyaniline	Nanostructure	Zhang et al. (2010)
Polyaniline	Quantum dots	Ma et al. (2003)
Ethylenediamine	Nanorods	Li et al. (2009)
Polyethylene glycol	Nanowires	Thongtem et al. (2009)
Polyethylene glycol	Nanocrystals	Abdolahzadeh et al. (2012)
Polyethylene glycol	Nanowires	Gai et al. (2008)
Polyethylene glycol	Nanoparticles	Thambidural et al. (2009)
Polystyrene	Quantum dots	Sajinovic et al. (2000)
Polystyrene	Hollow spheres	Wu et al. (2004)
Polystyrene	Nanoparticles	Antolini et al. (2005)
Polyvinyl alcohol	Nanocrystals	Wang et al. (2007)
Polyvinyl alcohol	Nanowires	Yao et al. (2005)
Polyvinyl alcohol	Nanowires	Gai et al. (2008)
Polyvinyl alcohol	Nano-dumbbells	Jing et al. (2010)
Perylene tetracarboxylic diimide	Nano-rose like	Song et al. (2014)
Polystyrene	Nanofiber	Chen et al. (2000)
Polystyrene	Nanowires	Yu et al. (2001)
Polystyrene	Nanoparticles	Du et al. (2002)
Polyvinyl-pyrrolidone	Nanoflowers	Song et al. (2012)
L-cysteine and ethanolamine	water lily-like nanocrystals	Xiong et al. (2010)
Polyvinyl acetate	Nanofibers	Xie et al. (1999)

Certain prerequisites have to be met for the film deposition in assistance of polymers. The soluble polymer must have proper interaction with the metal ions in order to avoid phase separation during the deposition process (Kalagi et al., 2010). In general, CdS nanostructures can be synthesized with different approach. CdS nanostructures and polymer may synthesize separately, followed by incorporating the preformed CdS nanostructures directly into polymer. Another approach is to anchor the Cd^{2+} ions into polymer supports, attachment of end-functionalized polymer to the nano-structure, and surface modification of the nanostructure (Zhou et al., 1999; Zhang et al., 2002; Zhang et al., 2004). Another approach is the attachment of functional groups to Cd^{2+}, followed by the monomer polymerization and S^{2-} addition to form CdS nanostructure. Thus, both CdS nanostructure and polymer are formed in situ from their precursors during the fabrication process. This method is found to be very effective to attach a large amount of CdS on polymer particles (Nair et al., 2005, Song et al., 2003, Sherman et al., 2005, Wu et al., 2005, Sooklal et al., 1998).

In this chapter, we are reporting herewith architecture of hierarchical nanostructured CdS via facile polymer assisted chemical bath deposition method using different polymers. Effect of polymers such as poly-vinylpyrrolidone (PVP), Polyethylenimine (PEI) and polyethylene glycol (PEG) on the growth of CdS crystals have been studied by conducting experiments with and without polymers.

11.2   MATERIALS AND METHODS

11.2.1   MATERIALS

All reagents and solvents were of analytical grade and were used as received without further purification. The cadmium sulfate ($CdSO_4 \cdot H_2O$) was used as cadmium (Cd) source and thiourea ($H_2N \cdot CS \cdot NH_2$) for sulfur (S) source. Liquor ammonia (NH_3) was used as a complexing agent. The polymers like polyvinylpyrrolidone (PVP), polyethylenimine (PEI) and polyethyl-ene glycol (PEG) were purchased from Sigma-Aldrich. Cadmium sulfate, thiourea and ammonia were purchased from s.d. fine chemicals

11.2.2   METHOD OF DEPOSITION

The chemical bath deposition method is perhaps the oldest known chemical method for the formation of thin films. This technique usually involves the simple immersion of a substrate into a solution containing both a metal salt and a chalcogenide precursor. As such, it has generally been dominated by the formation of sulfides and selenide films. CBD is a method in which controlled release of metal ions and chalcogenide ions take place that results in having control over grain size of thin films. The precipitation of metal chalcogenides in CBD occurs only when the ionic product exceeds the solubility product of metal chalcogenides (Chopra et al., 1982). Combinations of ions form nuclei on the substrate as well as in the solution results in precipitation. The film growth takes place via ion-by-ion condensation of the materials or by adsorption of colloidal particles from the solution onto the substrate. The complexing agents (like NH_3) help to control the reaction rate.

The preparative parameters like precursor concentration, deposition time and temperature were varied to yield good quality CdS thin films. The CdS films have been synthesized using two different wet chemical routes: (i) without polymer and (ii) with polymer. In the first route, aqueous ammonia (NH_4OH) was added into 1 M $CdSO_4$ solution to maintain 11 pH of the solution. The initial turbid solution is turned into a transparent, by adding excess ammonia. Then, 1 M thiourea was added into above solution. The CdS films were deposited by dipping the soda lime glass and fluorine-doped tin oxide (FTO) substrates into the above solution at 90°C for 10 min. The deposited films were rinsed in distilled water and dried at room temperature overnight. In the second route, polymer like PVP, PEI and PEG were added separately in definite proportions to the aqueous precursor solution. The concentration of polymers was kept constant at 1% in the final solution. Prior to the film deposition, the substrates were cleaned with detergent in distilled water and under ultrasonic cleaner, with acetone, alcohol and finally with distilled water. The films prepared by using different polymers such as PVP, PEI and PEG are denoted as CdS:PVP, CdS:PEI and CdS:PEG respectively, and those without polymer are denoted as CdS. After the deposition, the films were soaked in ethanol to remove stray organic molecules from the final deposits.

11.2.3   THIN FILM CHARACTERIZATION

The structural properties of the CdS thin films were studied using an X-ray diffractometer (Philips, PW 3710, Almelo, Holland) operated at

25 kV, 20 mA with CuKα radiation (1.5407 Å). The chemical composition and valence states of constituent elements were analyzed by X-ray Photoelectron Spectroscopy (XPS, Physical Electronics PHI 5400, USA) with monochromatic Mg-Kα (1254 eV) radiation source. Optical absorbance was measured using a UV-vis spectrophotometer (UV1800, Shimadzu, Japan). The surface morphology of the films was examined by SEM (Model JEOL-JSM-6360, Japan), operated at 20 kV. Field emission SEM (JSM-6701F model) was employed for closer insight into the CdS morphology. The thickness of the resulting CdS films was measured using surface profiler (Ambios XP-1).

11.2.4 PHOTO-ELECTROCHEMICAL SOLAR CELL

J-V characteristics were recorded using semiconductor characterization system (SCS-4200 Keithley, Germany) with two-electrode configuration under a halogen lamp (30 mW/cm²). Following cell configuration was used to record J-V plots: Glass/FTO/CdS/Na_2S-NaOH-S/G.

The CdS film (average area 1.2 cm²) and graphite rod (average area 1.2 cm²) were employed as the working and counter electrodes, respectively. The distance between the photoelectrode and counter electrode was 0.5 cm. The aqueous 1 M polysulphide (Na_2S+S+NaOH) was used as redox electrolyte. Measurements of the power output characteristics, and J–V plots were made at fixed intervals after waiting for sufficient time to equilibrate the system (both in the dark as well as under illumination). Plot of log I vs. V is used to calculate the junction ideality factor in dark. The schematic setup of PEC measurement is shown in inset of Fig. 11.10.

11.3 RESULT AND DISCUSSION

11.3.1 FILM FORMATION MECHANISM

The ionic reaction between Cd^{2+} and S^{2-} plays a vital role during the synthesis of CdS nanostructure. Since various precursors can be used for Cd^{2+} and S^{2-} ions. Chemical conditions differ for different chemical precursor.

Generally, cadmium chloride ($CdCl_2$), cadmium acetate ($Cd(CH_3CO_2)_2$) or cadmium sulfate ($CdSO_4.xH_2O$) is typical cadmium sources while sodium sulfide (Na_2S) is the commonly used sulfur source. But, Na_2S reacts rapidly with cadmium precursors; it becomes difficult to control the morphology of resulting CdS nanostructures. In this regard, thiourea ($SC(NH_2)_2$) is used as the sulfur source. Thiourea is an organo-sulfur compound, commonly employed as a source of sulfide. During the chemical process at moderate temperature, the S^{2-} anions could be gradually dissociated from $(NH_2)_2CS$, therefore, numerous complexed CdS clusters were generated. The formation of CdS thin films using NH_3 occurs via the following steps (Froment et al., 1995; Moualkia et al., 2009).

(i) Cadmium sulfate dissociation

$$CdSO_4 \rightarrow Cd^{2+} + SO^{4-} \qquad (1)$$

(ii) Ammonia dissociation

$$NH_4OH \rightarrow NH_3 + H_2O \qquad (2)$$

(iii) Complex formation

$$Cd^{2+} + 4NH_3 \rightarrow [Cd(NH_3)_4]^{+} \qquad (3)$$

(iv) Thiourea decomposition

$$CS(NH_2)_2 + 2OH^- \rightarrow S^{2-} + CH_2N_2 + 2H_2O \qquad (4)$$

(v) Formation of CdS

$$Cd^{2+} + S^{2-} \rightarrow CdS \qquad (5)$$

Apart from above mentioned reactions, several interactions exist between the CdS precursor and polymer during the synthesis of polymer assisted CdS nanostructure. These interactions include attachment of functional groups on polymer chains with nanostructure. Also, chelating and/or coordination contacts can be employed to restrain CdS nanostructure onto the surface of polymers. The interaction of CdS precursors with polymer is discussed in detail in Section 11.3.4.

11.3.2 STRUCTURAL PROPERTIES OF CDS NANOSTRUCTURE

X-ray diffraction (XRD) patterns were obtained for structural characterization. Figure 11.4 (a) shows the XRD pattern of with and without polymer mediated CdS nanostructure on the soda lime glass substrate. The films deposited did not show peaks related to elemental cadmium or sulfur or carbon. The XRD data confirm that all of the samples are crystalline and have the cubic structure of bulk CdS with lattice parameters matching those in the literature (JCPDS, 80–0019). No peaks of impurities are

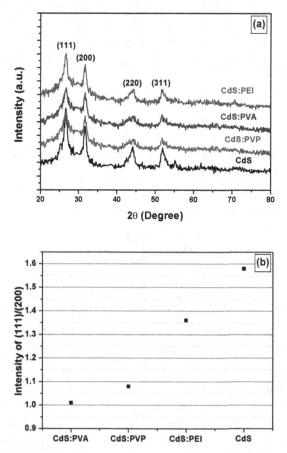

FIGURE 11.4 (a) X ray diffraction pattern of with and without polymer mediated CdS thin film. (b) Plot of the (100)/(002) intensity ratio of CdS, CdS:PVP, CdS:PEI and CdS:PEG thin film.

detected, indicating that the CdS nanostructures are pure and well crystalline. Four characteristics (111), (200), (220), (311) peaks appear at $2\theta =$ 26.45, 31.71, 44.13, and 51.87 degrees for various samples. The XRD patterns of film made with and without polymers are identical. This suggests that the organic polymers do not affect the crystallographic orientation of CdS. But, with the addition of polymers, a slight decrease of (111) intensity is observed. These results indicate that, the addition of polymers has a vital influence on CdS preferential growth orientation. As shown in Fig. 11.4 (b), the intensity ratio of (111)/(200) evidently decreasing with the addition of polymers. It was found earlier that the changing relative intensity of XRD peaks corresponds to a change in crystal shape. An unlike (111)/(200) ratio is indicative of the formation of different nanostructures with changing polymers (Lu et al., 2011; Song et al., 2014)

The lattice parameter 'a' is calculated using the following Eq. (6),

$$\frac{1}{d^2} = \frac{h^2 + k^2 + l^2}{a^2} \tag{6}$$

The mean values of a= 5.810 Å for CdS sample is in good agreement with the reported value a=5.811Å.

Further, using the breadth of (111) peak the average crystallite size is estimated using Scherrer's formula given below Eq. (7)

$$D_{(111)} = \frac{k\lambda}{\beta \cos\theta} \tag{7}$$

where $\lambda = 1.5406$Å, k is the dimensionless constant (0.95), β is the corrected broadening of the diffraction line measured at half of its maximum intensity (taken in radians by multiplying a factor of $\pi/360$) and D the diameter of crystallite and θ is diffraction angle. The calculated crystallite size is found to be 20 nm for (111) plane for CdS sample. Also, the broadened peaks indicate the nanocrystalline nature of the films.

Figure 11.5(a) shows the x-ray photoelectron spectroscopy-survey spectrum of the CdS:PEG samples. No peaks of other elements except cadmium (Cd), sulfur (S), carbon (C), and oxygen (O) are observed. The C and O peaks stem mainly from the atmospheric contamination due exposure of the sample to air. An unambiguous presence of the Cd3d

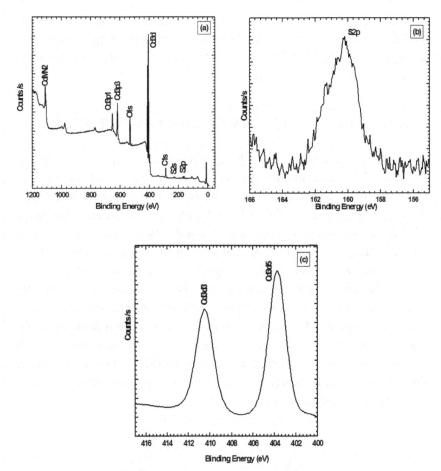

FIGURE 11.5 (a) X-ray photoelectron spectroscopy-survey spectrum of the CdS:PEG samples. (b) Narrow range scans for the S peak region of the CdS:PEG samples. (c) Narrow range scans for the Cd peak region of the CdS:PEG samples.

doublet signal at 404 eV and 419 eV clearly shows the formation of CdS. Figure 11.5 (b-c) depicts narrow range scans for the S and Cd peak region of the same samples. The binding energies obtained in the XPS analysis have been corrected taking into account the specimen charging and by referring to C1s at 284.88 eV. The two-peak structure in Cd 3d core level arises from the spin- orbit interaction with the Cd 3d5/2 peak position at 403.75 eV and the 3d3/2 at 410.48 eV. It is clear from the spectral graph that, Cd 3d exhibits narrow, well-defined feature for doublet structure.

This suggests that, specifically Cd atoms appear to bond to S atoms. The XPS binding energies of Cd 3d3 at 404.14 eV and the S 2p at 160.89 eV are indicative of the CdS chemistry. The peak originated at 1111.04 eV is due to the auger electron of Cd.

11.3.3   OPTICAL PROPERTIES OF CDS NANOSTRUCTURE

Due to the nanocrystalline thin films, UV-vis spectroscopy has become an effective tool in determining the size and optical properties. Figure 11.6 shows the room temperature optical absorption spectrum of the all CdS samples (CdS, CdS:PVP, CdS:PEI and CdS:PEG) recorded in the range of 450–700 nm without taking into account of scattering and reflection losses. Figure 11.6 presents UV-visible absorptions with onsets occurring at ~520 nm for all the CdS films, typical of the CdS bulk materials. It is notable that the absorption of polymer assisted CdS samples are slightly stronger than that of without polymer CdS sample. The slight improved absorption capacity of the polymer assisted CdS thin films could be attributed to positive modification of CdS nanostructure (Yang et al., 2013). A significant shift in the spectral photoresponse (520 nm) is observed for CdS:PEG sample. It clearly illustrates effective photon capturing in the visible region (Vanalakar et al., 2011)

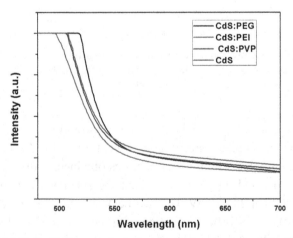

FIGURE 11.6   The room temperature optical absorption spectrum of the all CdS samples (CdS, CdS:PVP, CdS:PEI and CdS:PEG).

Absorbance coefficient (α) associated with the strong absorbance region of the films was calculated from the absorbance (A) and the film thickness (t) using the formula given below (equation 8)

$$\alpha = \frac{2.3026\ A}{t} \tag{8}$$

The absorption coefficient was analyzed using the following equation for near age optical absorption of the semiconductors. The Eq. (9) is

$$\alpha = \alpha_0 \frac{(h\upsilon - E_g)^n}{h\upsilon} \tag{9}$$

where, E_g is the optical energy band gap between the bottom of the conduction band and the top of the valence band, $h\upsilon$ is the photon energy and n is a constant whose value is ½ for direct transition and 2 for indirect transition. Figure 11.7 shows the variation of band gap energy (Eg) of

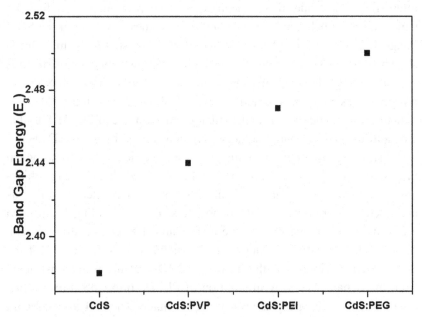

FIGURE 11.7   The variation of band gap energy (Eg) of CdS thin film with and without polymers.

CdS thin film with and without polymers. The slight increment in Eg is observed with CdS:PEG sample as compare with CdS sample.

11.3.4   SURFACE MORPHOLOGICAL PROPERTIES OF CDS NANOSTRUCTURE

Figure 11.8 shows the SEM images of CdS thin films prepared with and without polymers. The low magnification SEM images demonstrate that the CdS microstructures uniformly and compactly cover the entire substrate. The surface morphologies differ depending on the polymer used. The as synthesized CdS thin films are examined by SEM (Fig. 11.8(a) and (e)). The low magnification SEM image (Fig. 11.8(a)) reveals the film surface looks porous and the formation of CdS micro-flowers over the entire substrate. Also, there is not any overgrowth on the substrate without any void, pinholes or cracks and they cover the substrate well. The sizes of micro-flowers are observed around 5–6 μm. From high magnification SEM image of CdS thin film (Fig. 11.8(e)) it is clear that CdS micro-flower is made of nanopetals connected with each other forming nanoconduits. Such petals having ~60 nm thickness are seen in the SEM image Fig. 11.8(e). The SEM images of PVP assisted CdS thin film is shown in figure 11.8 b and f, from which cauliflower-like nanostructures with an average diameter of 10 to 12 μm were clearly observed. No other morphologies could be detected, indicating a high yield of these 3D nanostructures. It can be seen from the enlarged SEM image of Fig. 11.8(e) that the cauliflower-like nanostructures were constructed by many 2D sheets. The high-magnification SEM image shows the detailed morphological information of the nanosheets. It can be observed that the edge thickness of nanosheets was about 70 nm. Moreover, nanometer-sized porous architectures were formed in the nanosheets, as shown in Fig. 11.8(e). On the basis of the above results, the as-prepared flower-like structure can be generally classified as hierarchical structures. It is apparent that the as-prepared precipitates with a flower-like 3D structure were constructed from many nanosheets. With addition of PEI to form CdS, flowers like microstructure disappear as seen in SEM image. The PEI assisted CdS thin film shows the nanoflakes like structure over the entire substrate. High magnification SEM image (Fig. 11.8(g)) shows densely packed

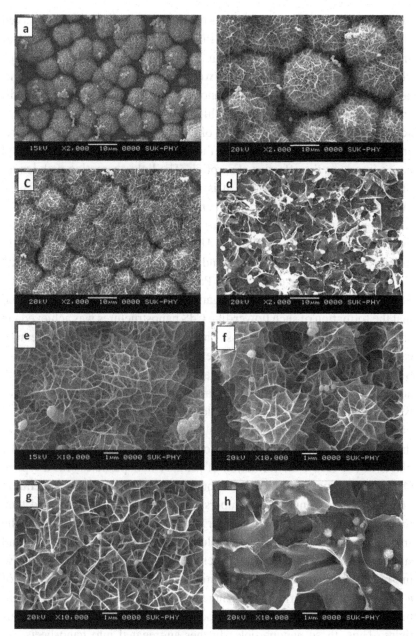

FIGURE 11.8 The low magnification SEM images of (a) without polymer CdS sample (b) CdS:PVP sample; (c) CdS:PEI sample and (d) CdS:PEG sample. The high magnification SEM images of (e) without polymer CdS sample (f) CdS:PVP sample (g) CdS:PEI sample and (h) CdS:PEG sample.

interconnected nanoflakes. The sizes of walls of nanoflakes were observed around 60 nm. With the assistant of PEG, the morphologies of the samples are gradually improved; the interconnected nanoflakes with nanoparticles like structure (Fig. 11.8(d) and (h)) were observed. Due to these interconnected nanowalls/nanopetals and nanoflakes, the PVP, PEI and PEG mediated CdS thin film sample provides high surface area for enhanced surface activities like efficient permeability of electrolytes into the inner structure, and light absorption. It also scatters the light multiply into photoanode and hence improves the effective absorption. The light absorption path length of photons can be increased as it is trapped in the nanoconduits. These mechanisms boost the PEC performance of CdS electrode.

The polymers play a key role as a blocking, capping, and orienting material in the growth of the CdS. The hydroxyl and carboxyl groups in the polymer ion attract Cd^{2+} ions and control the growth direction of the CdS structures (Kang et al., 2007, Pal et al., 2011). On the basis of the SEM observations and the investigations described above, we proposed that the formation of such flower-like CdS nanostructures was achieved via an "oriented attachment" process. The main evolving steps are schematically illustrated in Figure 11.9. In the initial stage, the aqueous solution of S^{2-}, Cd^{2+} and NH_3 leads to the formation of first nucleation seeds, which act as an initial nucleus along the particle growth. CdS tiny nanoparticles were formed at the early stage, followed by the oriented attachment of these building blocks into 2D nanosheets, and finally the arrangement of these sheets into 3D hierarchical flower-like nanostructures would take place. In our study, the formation of primary nanoparticles was a typical Ostwald ripening process (McNaught et al., 1997). At first, the generation of tiny crystalline nuclei in a supersaturated solution occurred and then followed by crystal growth. The larger particles grew at the cost of the smaller ones due to the different solubility between relatively larger and smaller particles. Following the general expression of Gibbs law, the driving force for crystal growth is to minimize the total surface free energy of a system. Certainly, oriented attachment would do so, and the oriented feature should further reduce the interface energy between primary nanoparticles. Thus, these nanoparticles further aggregated into nanosheets. As the reaction proceeded, the CdS nanosheets gradually evolved into flower-like structures through an oriented attachment and with the help of polymers (Cheng et al., 2008, Peng et al., 2011, Hongbing et al., 2011, Nguyen

Molecular structures of three kind of polymers

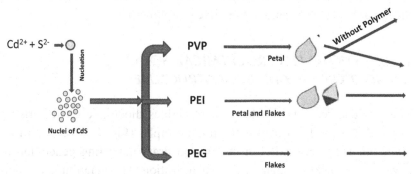

Possible formation process of CdS with different polymers

FIGURE 11.9   Molecular structures of PVP, PEI, PEG and Schematic representation of formation of CdS flower like morphologies.

et al., 2014). When the reaction was further progressed, more and more CdS nanosheets were assembled and eventually formed a flower-like hierarchical morphology. However, it was difficult to maintain the hierarchical flower-like structures in the growth and restructure of crystallite process when the deposition time was further prolonged. The obtained sample was mainly composed of sheet fragments. The detailed mechanism for the formation of porous CdS flower-like hierarchical structure is still under investigation by our group. Here is a working hypothesis that agreed well with the electron microscopy.

When PEI was used, in the initial stage, the aqueous solution of Cd^{2+}, S^{2-}, PEI and NH_3 leads to the formation of first nucleation seeds, which act as an initial nucleus along the particle growth. When the particle reaches a critical dimension, PEI absorbs on the small particles by the –OH bonds acting as a template for the formation of CdS flake-like nanostructures. Further, in the presence of PEG a relatively large difference is seen for the surface morphology of CdS thin films. Low magnification images revealed the formation of a large number of clusters of nanosheets (Fig. 11.8(d)). These nanosheets and clusters are relatively small in size and cover the entire substrate surface uniformly. Furthermore, high magnification images suggest that the relatively loosely packed nanosheets grow almost parallel to each other and provide abundant space (Fig. 11.8(h)). PEG molecules are aromatic hydrocarbon. These moieties generate nanoscale hydrophobic and hydrophilic regions. Therefore, the growth process was mediated by the nanoscale segregation of organic-rich and organic-lean domains on the growing CdS particles. Such nanoscale segregation is responsible for the noval flakes with nanoparticle-like morphology.

11.3.5 PHOTO-ELECTROCHEMICAL SOLAR CELL APPLICATION OF CDS NANOSTRUCTURE

The J-V characteristic in the dark resembles diode-like characteristics of the PEC cells fabricated with all the samples (Fig. 11.10). Upon illumination, J-V curve shifts in the IVth quadrant indicating generation of electricity, typical of solar cell characterization. The magnitude of short circuit current density (J_{SC}) was 1.27, 1.56, 1.82 and 2.14 mA/cm^2 for CdS, CdS:PVP, CdS:PEI and CdS:PEG samples, respectively. The corresponding open circuit voltage (V_{OC}) was found to be 435, 400, 453 and 496 mV. This observation reveals that the J_{SC} is increased while there is no subsequent rise in the V_{OC}. However, the shape of the J-V curves for all samples sag towards the origin. All samples show low value of fill factor (FF). We know the factors on which Jsc depends are:

1. thickness as maximum as possible for greater extends of absorptivity;
2. anti-refection coating to reduce reflection;
4. high diffusion length;
5. high surface area to increase the interfacial reaction sites;
6. optical properties of solar cell and the band gap;
7. morphological features (size, shape, grain boundary, inter connection of particles) that define energetic and kinetics at the electrode interface, and hence in influence the PEC efficiency of the solar cell;
8. electrode comprising a densely packed array can enhance PEC performance by virtue of the improvement in carrier transport mechanism and minimizing surface trap states.

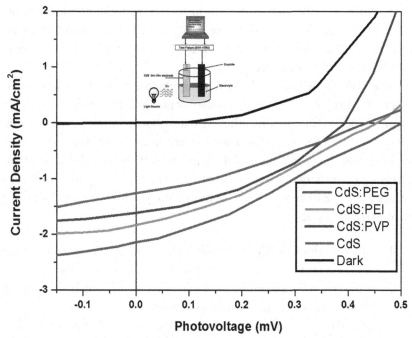

FIGURE 11.10 Photocurrent-density–voltage curves for sample CdS, CdS:PVP, CdS:PEI and CdS:PEG. (Inset shows the schematic setup of PEC measurement)

The factors on which Voc depends are:

1. Thickness must be optimum. For the low thickness, Voc increases but Jsc decreases. In any case, a cell should not be made unnecessary thick, not only in order to save on materials but for recombination;
2. Recombination;
3. Larger band gap means a larger Voc;
4. However, lower band gap gives greater Jsc which also increase Voc;
5. Dark or recombination current as low as possible.

The FF depends on the series resistance (R_S) and shunt resistance (R_{Sh}). The Rs is due to the resistance of the metal contacts, ohmic losses in the front surface of the cell, impurity concentration and junction depth. Ideally, the R_S should be 0Ω. The R_{Sh} represents the loss due to surface leakage along the edge of the cell or due to crystal defects. Ideally, the R_{Sh} should be infinite. For our CdS samples R_S varies slightly, from 205Ω to 100Ω, while R_{Sh} changes drastically, from 820Ω for CdS sample to 215Ω

for CdS:PEG sample. The variation in R_{Sh} seems to be dominant in our case. The consequences of decrease in R_{Sh} and FF along with conversion efficiency (η) are shown in Table 11.2. The highest η is of the order of 1.00% for CdS:PEG sample.

The observed values of J_{SC} in our samples are found to be larger than other CdS samples with compact planer or porous morphologies. The photocurrent depends upon how efficiently the photogenerated carriers in the semiconductors are harvested. Photocurrent results from two main collection mechanisms, separation of the carriers in the space charge field and diffusion of carriers towards the interface. The hike in J_{SC} in our samples seems to be due to spongy ball like morphology, which induces following light harvesting phenomena:

1. Nanoconduits with interconnected nanopetals or nanoflakes increase surface area in contact with the redox electrolyte.
2. Effective light absorption by the way of its trapping and scattering in the nanoconduits. It increases absorption path length of light and thereby interaction between light and CdS nanocrystallites increases. Hence, the optical absorption and energy harvesting efficiency of CdS spongy balls are enhanced.
3. The nanowalls of 70 nm thickness act essentially as space charge (depletion) region, which absorb and create electron-hole pair and separate them effectively due to the built-in potentials at nanowalls-electrolyte interface.
4. The interconnected nanowalls serve as direct pathways for photogenrated electrons and reduce its loss.

TABLE 11.2 The Values of Solar Cell Parameters for CdS, CdS:PVP, CdS:PEI and CdS:PEG Thin Films

Sample	JSC (mA/cm2)	VOC (Volt)	Fill Factor	Efficiency (%)
CdS	1.27	435	0.30	0.55
CdS:PVP	1.56	400	0.31	0.66
CdS:PEI	1.82	453	0.30	0.82
CdS:PEG	2.14	496	0.29	1.00

The ideality factor 'n_d' of prepared CdS films is determined from following diode equation (Eq. (10)) as,

$$I = I_O \left(e^{qV/ndkT} \right) - 1 \tag{10}$$

where, Io is the reverse saturation current, V is forward bias voltage, k is Boltzmann's constant, T is ambient temperature in Kelvin and n_d is an ideality factor. The ideality factor is determined under forward bias and is normally found to be in between 1 to 2 depending up on the relation between diffusion current and recombination current. When diffusion current is more than recombination current then ideality factor becomes 1 and it becomes 2 in opposite case. The ideality factor was found to be 1.86, 1.72, 2.06 and 2.08 for the sample CdS, CdS:PVP, CdS:PEI and CdS:PEG, respectively.

The ideality factor is a fitting parameter that describes how closely the diode's behavior matches the behavior predicted by theory. The ideality factor is determined under forward bias and is normally found to be in between 1 to 2 depending upon the relation between diffusion current and recombination current. The ideality factor becomes 1 when the p-n junction of the diode is an infinite plane and no recombination occurs within the space-charge region. When recombination current is more than diffusion current then ideality factor becomes 2. The ideality factor 'nd' of prepared CdS films is determined from following diode equation as,

$$I = I_O \left(e^{qV/ndkT} \right) - 1$$

where, Io is the reverse saturation current, V is forward bias voltage, k is Boltzmann's constant, T is ambient temperature in Kelvin and nd is an ideality factor. The higher value of nd is indicative of the series resistance effect, surface states and the charge carrier recombination at the semiconductor–electrolyte interface. These factors reduce the ideality of devices.

11.4 CONCLUSION

The successful growth of CdS by polymer assisted chemical bath deposition method suggests that this is a useful alternative to the growth of high-quality metal chalcogenide films. The key features in polymer assisted chemical bath deposition are the de-polymerization of the polymer and the protection of the metal until the last moments of deposition when the metal chalcogenide film is formed. In addition this bottom-up growth method provides a unique capability in conformal coating of nanostructures. In summary, CdS composed of uniform interconnected nanowalls network have been successfully prepared via a polymer-assisted wet chemical reaction system at low temperature and ambient atmosphere. This method is simple and suitable for the preparation of large surface area thin film in a one step process. The as deposited CdS film showed a cubical crystal structure. The well-covered porous structure with the interconnected nanowalls/nanopetals network morphology leading to high surface area is observed by SEM studies. This structure is a good probable way for PEC application

ACKNOWLEDGMENT

This work was supported by the Human Resources Development of the Korea Institute of Energy Technology Evaluation and Planning (KETEP) grant funded by the Korea government Ministry of Knowledge Economy (No.: 20124010203180). The author SAV is grateful to DST-SERB, New Delhi for financial support (SR/FTP/PS-083/2012).

KEYWORDS

- cadmium sulfide
- characterization
- chemical bath deposition
- morphology
- nanostructure
- PEG
- PEI
- polymer
- PVP
- solar cell

REFERENCES

1. Abdolahzadeh Z. A.; Ghodsi F.E., Growth, characterization and studying of sol–gel derived CdS nanoscrystalline thin films incorporated in polyethyleneglycol: Effects of post-heat treatment, *Sol. Energy Mater. Sol. Cells* 105, 249–262 (2012).

2. Antolini F.; Pentimalli M.; Luccio T. D.; Terzi R.; Schioppa M.; Re M.; Mirenghi L.; Tapfer L., Structural characterization of CdS nanoparticles grown in polystyrene matrix by thermolytic synthesis, *Mater. Lett.* 59, 3181–3187 (2005).

3. Bhirud A.; Chaudhari N.; Nikam L.; Sonawane L.; Patil K.; Baeg J. O.; Kale B. B., Surfactant tunable hierarchical nanostructures of $CdIn_2S_4$ and their photohydrogen production under solar light *Int. J. Hydrogen Energy* 36, 11628–11639 (2011).

4. Burrell A. K.; McCleskey T. M.; Jia Q. X., Polymer assisted deposition, *Chem. Commun.* 31, 1271–1277 (2008).

5. Chen M.; Xie Y.; Qiao Z.; Zhua Y.; Qiana Y., Synthesis of short CdS nanofiber/poly(styrene-alt-maleic anhydride) composites using γ-irradiation *J. Mater. Chem.* 10, 329–332 (2000).

6. Cheng X.; Zhao Q.; Yang Y.; Chin S.; Li K.Y., A facile method to prepare CdS/polystyrene composite particles, *J. Colloid Interface Sci.* 326, 121–128 (2008).

7. Chopra K. L.; Kainthla R. C.; Pandya D. K.; Thakoor A. P., Chemical Solution Deposition of Inorganic Films *Phys. Thin Films.* 12, 167–235 (1982).

8. Chopra K. L.; Kainthla R. C.; Pandya D. K.; Thakoor A. P., *Physics of Thin Films*, Academic Press: New York, 11–167 (1982).

9. Cruz J. S.; Pe'rez R. C.; Delgado G. T.; Angel O. Z., CdS thin films doped with metal-organic salts using chemical bath deposition, *Thin Solid Films* 518, 1791–1795 (2010).

10. Dhage S. R.; Colorado H. A.; Hahn T., Morphological variations in cadmium sulfide nanocrystals without phase transformation, *Nanoscale Res. Lett.* 6, 420–425 (2011).

11. Du H.; Xu G. Q.; Chin W. S.; Huang L.; Ji W., Synthesis, Characterization, and Nonlinear Optical Properties of Hybridized CdS–Polystyrene Nanocomposites *Chem. Mater.* 14, 4473–4479 (2002).

12. Froment M.; Lincot D., Phase formation processes in solution at the atomic level: Metal chalcogenide semiconductors *Electrochim. Acta.* 40, 1293–1303 (1995).

13. Gai H.; Wu Y.; Wang Z.; Lili W.; Shi Y.; Jing M.; Zou K., Polymer-assisted solvothermal growth of CdS nanowires *Poly. Bull.* 61, 435–441 (2008).

14. Gao M. R.; Xu Y. F.; Jiang J.; Shu-Hong Yu S. H., Nanostructured metal chalcogenides: synthesis, modification, and applications in energy conversion and storage devices, Chem. Soc. Rev., 42, 2986–3017 (2013).

15. Hanemann T.; Szabo D. V., Polymer-Nanoparticle Composites: From Synthesis to Modern Applications *Mater. Lett.* 3, 3468–3517 (2010).

16. Hodes G., *Chemical Solution Deposition of Semiconductor Films*, Marcel Dekker: New York, 54–78 (2003).

17. Hongbing L.; Wang S.; Li Z.; Li J.; Donga B.; Zuxun X., Hierarchical ZnO microarchitectures assembled by ultrathin nanosheets: hydrothermal synthesis and enhanced photocatalytic activity, *J. Mater. Chem.* 21, 4228–4234 (2011).

18. Inamdar A. I.; Mujawar S. H.; Ganesan V.; Patil P. S.; Surfactant-mediated growth of nanostructured zinc oxide thin films via electrodeposition and their photoelectrochemical performance, *Nanotechnology* 19, 325706–325713 (2008).

19. Jing M.; Gai H.; Wang Z.; Jiang K.; Wu L.; Wu Y., Poly (vinyl alcohol)-assisted solvothermal growth of CdS dumbbells and necklaces *Polym. Bull.* 64, 413–419 (2010).

20. Kalagi S.S.; Dalavi D.S.; Pawar R.C.; Tarwal N.L.; Mali S.S.; Patil P.S., Polymer assisted deposition of electrochromic tungsten oxide thin films *J. Alloys Compd.* 493, 335–339 (2010).

21. Kang C. C.; Lai C. W.; Peng H. C.; Shyue J. J.; Chou P. T., Surfactant- and temperature-controlled CdS nanowire formation, *Small* 3, 1882–1885 (2007).

22. Kariper A.; Guneri E.; Gode F.; Gumus C.; Ozpozan T., The structural, electrical and optical properties of CdS thin films as a function of pH, *Mater. Chem. Phys.* 129, 183–188 (2011).

23. Khallaf H.; Oladeji I. O.; Chow L., Optimization of chemical bath deposited CdS thin films using nitrilotriacetic acid as a complexing agent, *Thin Solid Films* 516, 5967–5973 (2008).

24. Kippeny T.; Swafford L.A.; Rosenthal S.J., Semiconductor Nanocrystals: A Powerful Visual Aid for Introducing the Particle in a Box, Journal of Chem. Educ. 79, 1094–1100 (2002).

25. Lai C. H.; Lua M. Y.; Chen L. H., Metal sulfide nanostructures: synthesis, properties and applications in energy conversion and storage *J. Mater. Chem.* 22, 19–30 (2012).

26. Lejmi N.; Savadogo S., The effect of heteropolyacids and isopolyacids on the properties of chemically bath deposited CdS thin films, *Sol. Energy Mater. Sol. Cells* 70, 71–83 (2001).

27. Li Y. X.; Hu Y. F.; Peng S. Q.; Lu G. X.; Li S. B., Synthesis of CdS nanorods by an ethylenediamine assisted hydrothermal method for photocatalytic hydrogen evolution. *J Phys Chem C* 113, 9352–9358 (2009).

28. Li Y. X.; Hu Y. F.; Peng S. Q.; Lu G. X.; Li S. B., Synthesis of CdS nanorods by an ethylenediamine assisted hydrothermal method for photocatalytic hydrogen evolution. *J Phys Chem C* 113, 9352–9358 (2009).

29. Liang Y.; Wu Y.; Feng D.; Tsai S. T.; Son H. J.; Li G.; Yu L., Development of new semiconducting polymers for high performance solar cells. *J Am Chem Soc* 131, 56–57(2008).

30. Lincot D.; Froment M.; Cachet H.; Alkire R. C.; Kolb D. M., *Advances in Electrochemical Science Engineering*, Wiley-VCH: New York, 6 (1998).

31. Lokhande C. D., Chemical deposition of metal chalcogenide thin films, *Mater. Chem. Phys.* 27, 1–43 (1991).

32. Lu, H.; Wang S.; Zhao L.; Li J.; Dong B.; Xu Z. Hierarchical ZnO microarchitectures assembled by ultrathin nanosheets: hydrothermal synthesis and enhanced photocatalytic activity *J. Mater. Chem.* 21, 4228–4234 (2011).

33. Ma X.; Shi W., Investigation of quantum size effect of laser induced CdS quantum dots in sulfonic group polyaniline (SPAn) film, *Microelectronics Engineering* 66, 153–158 (2003).

34. Mai Y.-W.; Tjong S. C., *Physical Properties and Applications of Polymer Nanocomposites*, Woodhead publishing Ltd, 3–13 (2010).

35. Mane R.S.; Lokhande C.D., Chemical deposition method for metal chalcogenide thin films, *Mater. Chem. Phys.* 65, 1–31 (2000).

36. McNaught D.; Wilkinson A.; Nic M.; Jirat J.; Kosata B., *IUPAC Compendium of Chemical Terminology*, 2nd ed. The Gold Book, Blackwell Scientific Publications:Oxford 60–78 (1997).

37. Moss T.S., Optical Properties of Semiconductors, Academic Press: New York 71–84 (1974).
38. Moualkia H.; Hariech S.; Aida M. S.; Attaf N.; Laifa E. L., Growth and physical properties of CdS thin films prepared by chemical bath deposition *J. Phys. D: Appl. Phys.* 42, 135404–135410 (2009).
39. Nair P. K.; Nair MTS; Garcia V. M.; Arenas O. L.; Pena Y.; Castillo A.; Ayala I. T.; Gomez-daza O.; Sanchez A.; Campos J., Hu H.; Suarez R., Rincon M. E., Semiconductor thin films by chemical bath deposition for solar energy related applications, *Sol. Energy Mater. Sol. Cells* 52, 313–324 (1998).
40. Nair P. S.; Radhakrishnan T.; Revaprasadu N.; Kolawole G. A.; Luyt A. S.; Djokovic V., Polystyrene-co-maleic acid/CdS nanocomposites: Preparation and properties *J. Phys. Chem. Sol.* 66, 1302–1306 (2005).
41. Nguyen T. K.; Kim S. W.; Thuan D.V.; Yoo D. H.; Kimb E. J.; Hahn S. H., Hydrothermally controlled ZnO nanosheet self-assembled hollow spheres/hierarchical aggregates and their photocatalytic activities, *Cryst. Eng. Comm.* 16, 1344–1350 (2014).
42. Okada A.; Kawasumi M.; Kurauchi T.; Kamigaito O., Polymer Preparation *Am. Chem. Soc. Div. Polym. Chem.* 28, 447–449 (1987).
43. Pal K.; Maiti U. N.; Majumder T. P.; Debnath S. C., A facile strategy for the fabrication of uniform CdS nanowires with high yield and its controlled morphological growth with the assistance of PEG in hydrothermal route, *Appl. Surf. Sci.* 258, 163–168 (2011).
44. Peng S.; Zhao W.; Cao Y.; Guan Y.; Sun Y.; Lu G., Porous SnO_2 hierarchical nanosheets: hydrothermal preparation, growth mechanism, and gas sensing properties, *Cryst. Eng. Comm.* 13, 3718–3723 (2011).
45. Rezaul K. M.; Woo L. H., Conducting polyaniline-titanium dioxide nanocomposites prepared by inverted emulsion polymerization, *Polym. Compos.* 31, 83–88 (2010).
46. Roy P.; Srivastava S.K., A new approach towards the growth of cadmium sulphide thin film by CBD method and its characterization, *Mater. Chem. Phys.* 95, 235–241(2006).
47. Sajinovic D.; Saponjic Z. V.; Cvjeticanin N.; Cincovic M. M.; Nedeljkovic J. M., Synthesis and characterization of CdS quantum dots-polystyrene composite Chem. Phys. Lett. 329, 168–172 (2000).
48. Savadogo O., Chemically and electrochemically deposited thin films for solar energy materials *Sol. Energy Mater. Sol. Cells* 52, 361–388 (1998).
49. Saxena T.; Rumyantsev S. L.; Dutta P. S.; Shur M., CdS based novel photo-impedance light sensor, *Semicond. Sci. Technol.* 29, 025002–06 (2014).
50. Senthil K.; Mangalaraj D.; Narayandass S.K., Structural and optical properties of CdS thin films, *Appl. Surf. Sci.* 169, 476–479 (2001).
51. Shang M.; Wang W. Z.; Sun S. M.; Ren J.; Zhou L.; Zhang L. Efficient visible light-induced photocatalytic degradation of contaminant by spindle-like PANI/$BiVO_4$ *J Phys Chem C* 113, 20228–32 (2009).
52. Sherman Jr. R.L.; Ford W.T., Semiconductor Nanoparticle/Polystyrene Latex Composite Materials *Langmuir* 21, 5218–5222 (2005).
53. Sone E. D.; Zubarev E. R.; Stupp S. I., Semiconductor Nanohelices Templated by Supramolecular Ribbons *Angew. Chem., Int. Ed.,* 41, 1705–1709 (2002).

54. Song C.; Gu G.; Lin Y.; Wang H.; Guo Y.; Fu X.; Hu Z., Preparation and characterization of CdS hollow spheres *Mater. Res. Bull.* 38, 917–924 (2003).

55. Song G.; Zhang H.; Li J.; Peng Z.; Li X.; Chen L., Poly(vinyl-pyrrolidone) assisted hydrothermal synthesis of flower-like CdS nanorings, *Polym. Bull.* 68, 2061–2069 (2012).

56. Song J.; Tian Q.; Gao J.; Wu H.; Chen Y.; Li X., Controlled preparation of CdS nanoparticle arrays in amphiphilic perylene tetracarboxylic diimides: organization, electron-transfer and semiconducting properties, *Cryst. Eng. Comm.* 16, 1277–1281 (2014).

57. Sooklal K.; Hanus L. H.; Ploehn H. J.; Murphy C. J., A Blue-Emitting CdS/Dendrimer Nanocomposite *Adv. Mater.* 10, 1083–1087 (1998).

58. Tang D.; Li H.; Niessner R.; Xu M.; Gao Z.; Knopp D., Multiplexed electrochemical immunoassay of biomarkers using metal sulfide quantum dot nanolabels and trifunctionalized magnetic beads, *Biosens. Bioelectron.* 46, 37–43 (2013).

59. Thambidurai M.; Murugan N.; Muthukumarasamy N.; Vasantha S.; Balasundaraprabhu R.; Agilan S., Preparation and characterization of nanocrystalline CdS thin film, *Chalcog. Lett.* 6, 171–179 (2009).

60. Thongtem T.; Phuruangrat A.; Thongtem S., Solvothermal synthesis of CdS nanowires templated by polyethylene glycol, *Ceram. Int.* 35, 2817–2822 (2009).

61. Trindade T.; O'Brien P.; and Pickett N. L., Nanocrystalline Semiconductors: Synthesis, Properties, and Perspectives *Chem. Mater.* 13, 3843–3858 (2001).

62. Vanalakar S. A.; Mali S. S.; Suryavanshi M. P.; Patil P. S., Quantum size effect in chemosynthesized nanostructured CdS thin films, *Dig J Nanomater Bios.* 5, 805–810 (2010).

63. Vanalakar S. A.; Pawar R. C.; Suryawanshi M. P.; Mali S. S.; Dalavi D. S.; Moholkar A. V.; Sim K.U.; Kown Y. B.; Kim J. H.; Patil P. S., Low temperature aqueous chemical synthesis of CdS sensitized ZnO nanorods Materials Letters 65, 548–551 (2011).

64. Vanalakar S.A.; Mali S.S.; Pawar R.C.; Tarwal N.L.; Moholkar A.V.; Kim Jin A.; Kwon Ye-bin; Kim J.H.; Patil P.S., Synthesis of cadmium sulfide spongy balls with nanoconduits for effective light harvesting *Electrochim. Acta* 56, 2762–2768 (2011).

65. Vanalakar S.A.; Pawar R.C.; Suryawanshi M.P.; Mali S.S.; Dalavi D.S.; Moholkar A.V.; Sim K.U.; Kown Y.B.; Kim J.H.; Patil P.S., Low temperature aqueous chemical synthesis of CdS sensitized ZnO nanorods, *Mater. Lett.* 65, 548–551 (2011).

66. Wang H.; Fang P.; Chen Z.; Wang S. H., Synthesis and characterization of CdS/PVA nanocomposite films, *Appl. Surf. Sci.* 253, 8495–8499 (2007).

67. Wiberg E.; Holleman A.F., *Inorganic Chemistry*, Academic Press: New York, 522–600 (2001).

68. Wu D.; Ge X.; Zhang Z.; Wang M.Z.; Zhang S.L., Novel one-step route for synthesizing CdS/polystyrene nanocomposite hollow spheres, *Langmuir* 20, 5192–5195 (2004).

69. Wu X. C.; Bittner A. M.; Kern K., Synthesis, Photoluminescence, and Adsorption of CdS/Dendrimer Nanocomposites *J. Phys. Chem. B* 109, 230–239 (2005).

70. Xiao M. W.; Wang L. S.; Wu Y. D.; Huang X. J.; Dang Z., Preparation and Characterization of CdS nanoparticles decorated into titanate nanotubes and their photocatalytic properties, Nanotechnology 19, 015706–11 (2008).

71. Xie Y.; Qiao Z.; Chen M.; Zhu Y.; Qian Y. T., Spherical assemblies of CdS nanofibers in poly(vinyl acetate) by g-irradiation *Nanostruct. Mater.* 11, 1165–1169 (1999).

72. Xiong S.; Xi B.; Qian Y., CdS hierarchical nanostructures with tunable morphologies: preparation and photocatalytic properties. *J Phys Chem C* 114, 14029–14035 (2010).

73. Yang C.; Li M.; Zhang W. H.; Li C., Controlled growth, properties, and application of CdS branched nanorod arrays on transparent conducting oxide substrate, *Sol. Energy Mater. Sol. Cells* 115, 100–107 (2013).

74. Yang X. H.; Wu Q. S.; Lia L.; Ding Y. P.; Zhang G. X., Controlled synthesis of the semiconductor CdS quasi-nanospheres, nanoshuttles, nanowires and nanotubes by the reverse micelle systems with different surfactants, *Colloids and Surfaces A: Physicochem. Eng. Aspects* 264, 172–178 (2005).

75. Yao J. X.; Zhao G. L.; Wang D.; Han G. R., Solvothermal synthesis and characterization of CdS nanowire/PVA composite films, *Mater. Lett.* 59, 3652–3655 (2005).

76. Yu S. H.; Yang J.; Han Z. H.; Zhou Y.; Yang R. Y.; Qian Y. T.; Zhang Y. H., Controllable synthesis of nanocrystalline CdS with different morphologies and particle sizes by a novel solvothermal process, *J. Mater. Chem.* 9, 1283–1287 (1999).

77. Yu S. H.; Yoshimura M.; Moreno J. M. C.; Fujiwara T.; Fujino T.; Teranishi R., In Situ Fabrication and Optical Properties of a Novel Polystyrene/Semiconductor Nanocomposite Embedded with CdS Nanowires by a Soft Solution Processing Route *Langmuir* 17, 1700–1707 (2001).

78. Zhang H.; Zhu Y. F. Significant visible photoactivity and antiphotocorrosion performance of CdS photocatalysts after monolayer polyaniline hybridization. *J Phys Chem C* 114, 5822–5826 (2010).

79. Zhang J.; Coombs N.; Kumacheva E., A New Approach to Hybrid Nanocomposite Materials with Periodic Structures *J. Am. Chem. Soc.*, 124, 14512–14513 (2002).

80. Zhang M.; Drechsler M.; Müller H. E., Template-Controlled Synthesis of Wire-Like Cadmium Sulfide Nanoparticle Assemblies within Core–Shell Cylindrical Polymer Brushes *Chem. Mater.* 16, 537–543 (2004).

81. Zhang S.; Chen Q.; Wang Y.; Liejin G., Synthesis and photoactivity of CdS photocatalysts modified by polypyrrole, *Int J Hydrogen Energy* 37, 13030–13036 (2012).

82. Zhou Y.; Yu S.; Wang C.; Zhu Y.; Chen Z.; Li X., A novel in situ simultaneous polymerization–hydrolysis technique for fabrication of polyacrylamide–semiconductor MS(M = Cd, Zn, Pb) nanocomposites *Chem. Commun.*, 13, 1229–1230 (1999).

83. Zuo Z.; Li Y., Functional polymers for photovoltaic devices, *Polym. Bull.* 68, 1425–1467 (2012).

CHAPTER 12

SYNTHESIS AND CHARACTERIZATION CHITOSAN-STARCH CROSSLINKED BEADS

VIRPAL SINGH

Department of Chemical Technology, Sant Longowal Institute of Engineering and Technology, Longowal, Sangrur, Punjab–148106, India, E-mail: singh_veer_pal@rediffmail.com

CONTENTS

12.1 Introduction... 205
12.2 Material and Methods 207
12.3 Experimental 207
12.4 Result and Discussion 210
12.5 Conclusion 215
Keywords 215
References... 216

12.1 INTRODUCTION

Generally natural or semi synthetic polymers are preferred as vehicle for drug delivery as they are biodegradable, non-toxic as well as biocompatible. Chitosan is a modified natural carbohydrate polymer prepared by the partial N-deacetylation of chitin. Chitosan is the deacetylated derivative of chitin, which is a water insoluble polymer. Chitosan dissolves readily in dilute solutions of most of organic acids including malic, acetic, tartaric, glycolic, citric and ascorbic acid solution. Chitosan, (1, 4)-(2amino-2-deoxy-β-D-glucan) is a natural biopolymer, obtained by a partial deacetylation of chitin. Chitin is a biopolymer abundantly available in nature and is found in the exoskeleton of crustaceans, in fungal cell walls and in other biological materials. Chitosan is inert, hydrophilic, and biocompatible and biodegradable (Dutta et al., 2004; Rinaudo et al., 1999; Rinaudo, 2006). The use of chitosan in pharmaceutical industry is still very limited because of is its high cost, poor mechanical strength, fast dissolution in the stomach for oral administration, and limited capacity to controlled drug release. Hence, other biodegradable materials such as pectin, Gaur gum, sodium alginate, and starch, etc., are used to reduce the cost (Zhang and Sun, 2004). In these materials, Cornstarch is low cost and easily available with an extra advantage of its compatibility with chitosan. Starch enhances the release of drug when added to controlled release formulations (Elviraa et al., 2002). Starch is mainly composed of two homopolymers of D-glucose; amylase and amylopectin. Amylase is mostly linear a-D (1, 4)-glucan and amylopectin has a backbone structure as amylose but with many a-1, 6 linked branch points (Atyabi et al., 2006). Starch is water swellable excipient in nature. The main differences between starch and chitosan are the glucoside linkage: α (1,4) for starch and β (1,4) for chitosan and, the hydroxyl group of the second carbon is replaced by the amine group.

Chitosan beads are usually prepared by precipitation, anionic cross-linking, chemical cross-linking, and thermal cross-linking, co-acervation and emulsification ionic-gelation methods (Srinatha et al., 2008). Problems of beads are such as low strength, high dispersion and solubility. Many processes have been used for crosslinking of chitosan, but easiest and cheapest way is formation of Schiff base between the aldehyde functional group glutaraldehyde (crosslinking agent) and amine group of chitosan. Their solubility/dispersion can be reduced by the use of cross linkage agents. Chitosan (Shu and Zhu, 2000; 2002) and chitosan-amino acid beads (Rani et al., 2010) are also used for in vitro release of drug. Chemical cross-linking improves the mechanical strength, thermal stability and swelling properties of the beads. Chitosan beads were hard by treatment of cross-linking agents such as glutaraldehyde, glyoxal and glycol. Chemical and mechanical behavior of beads are enhanced by crosslinking with bifunctional compounds such glutaraldehyde (Jameela and Jayakrishnan, 1995) and glyoxal etc. Chitosan beads and microgranules prepared from chitosan (Gupta and Kumar, 2000). IPN were the beads that of chitosan, chitosan-glycine and chitosan-glutamic acid cross-linked with glutaraldehyde (Rani et al., 2011). Synthesis of calcium alginate-chitosan beads, calcium from calcium alginate and chitosan and used to study the swelling behavior and the in vitro release of the antihypertensive drug verapamil hydrochloride. Calcium–alginate beads, chitosan-coated alginate beads and alginate–chitosan mixed beads were characterized by scanning electron microscopy (Pasparakis and Bouropoulos, 2006).

12.2 MATERIAL AND METHODS

12.2.1 MATERIALS

Chitosan is supplied by Fluka Bio Chemica and starch is procured from Himedia (India). Acetic acid (99.5%) is purchased from Merck (Germany) and Glutaraldehyde is procured Central Drug House (P) Ltd New Delhi. Sodium hexa meta phosphate is purchased from the Pioneer chemical company New Delhi. Chlorpheniramine maleate (CPM) [$C_{16}H_{19}ClN_2C_4H_4O_4$] a drug is obtained as gift sample from Japson Pharmaceutical Ltd Sangrur, India. For the preparation of solutions, double distilled water is used.

12.3 EXPERIMENTAL

12.3.1 SAMPLE PREPARATION

12.3.1.1 Preparation of Chitosan-Starch Beads

To synthesis of beads, a known quantity of chitosan is dissolved in 20 mL of 2% acetic acid solution at $25\pm2°C$ with continuous stirring for three hours. The starch solution is prepared by dissolving known weight of starch in 20 mL of water at 85°C while stirring for 20 minutes followed by natural cooling to room temperature. The solutions of chitosan and starch are mixed together and kept for 24 hours at room temperature

(25°C) in order to get a clear solution. The 0.2 g of CPM drug is added to the resultant solution containing chitosan and starch and mixed thoroughly. Then 20 mL of SHMP was added to the mixture of chitosan, starch and CPM. Then 20 mL of GA was added to it. This homogeneous mixture is extruded in the form of droplets using a 0.56 mm diameter syringe into alkali-methanol solution (1:20 w/w) under stirring conditions. The beads are washed with water and resultant beads are allowed to react with 20 mL of SHMP (7.5%, 10%, 12.5%, 15%, and 17.5%) for 20 minutes at room temperature (25°C). The beads are washed with distilled water and obtained beads are subjected to further crosslinking with 20 mL glutaraldehyde at 60°C for 10 minutes. These beads are washed with distilled water for removing unreacted glutaraldehyde from beads. The double cross-linked beads are dried at 40°C for 24 hours. The synthesized undried and dried beads show in Fig. 12.1.

The formulations of different synthesized beads are presented in Table 12.1.

12.3.2 TEXTURE ANALYSIS OF BEADS

The hardness Harness of chitosan-starch cross-linked beads was measured by Stable Micro System TA XT 2i Texture Analyzer. The instrument conditions were fixed as following for measuring hardness. The probe was fixed at height 10 mm from the sample plate. Text parameters were selected as pre-test speed (2 mm/s), test speed (1 /mm/s), distance (2mm), trigger

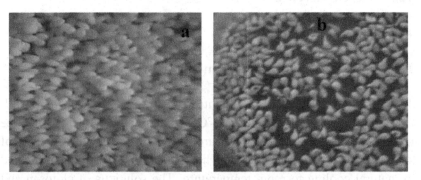

FIGURE 12.1   Wet synthesized beads and dry beads.

TABLE 12.1   Composition of Synthesis of Chitosan-Starch Beads

S.No	Chitosan (%)	Starch (%)	SHMP (12.5%) mL	Glutaraldehyde (25%) mL
1	50	50	20	20
2	60	40	20	20
3	70	30	20	20
4	80	20	20	20
5	90	10	20	20

force (5 g), and probe (p/4) at 50% strain (compression). The hardness was measured in terms of Kg.

12.3.3   SWELLING STUDIES

Swelling of profitable for polymer matrix depends upon the particular swelling media, which means a specific pH and temperature. Swelling studies are performed in solutions of pH 2.2 and pH 7.4 for understanding the molecular transport of liquid into cross-linked beads. A definite amount of dry sample is immersed in solutions of pH 2.2 and 7.4 at 37°C. At different time intervals beads is taken out and blotted off in between tissue paper (without pressing hard) to remove the surface adhered solution. The final weight of the samples is noted and percentage of swelling is calculated as (Barreiro-Iglesias et al., 2005; Freiberg and Zhu, 2004; Pasparakis and Bouropoulos, 2006; Peppas and Colombo, 1997; Vlachou et al., 2001):

$$S = \frac{W_s - W_d}{W_d} \times 100 \tag{1}$$

where W_s is the weight of swelled sample and W_d is the weight of the dry sample.

12.3.4   SCANNING ELECTRON MICROSCOPY (SEM)

The surface morphology of the chitosan-starch beads and cross-linked films is studied with the help of SEM (Rani et al., 2011). SEM analysis is made on JOEL scanning electron spectroscopy machine. Before focusing electron beam on the samples, the samples were gold-sputtered in order to make them conducting.

12.3.5 FTIR SPECTROSCOPY

FTIR spectra of chitosan-starch and cross-linked beads (Rani et al., 2011) were recorded by Perkin Elmer. FTIR spectrophotometer using KBr pellets. FTIR spectra of the samples were taken in the range of 400–4000 cm^{-1} with a resolution of 2cm^{-1}.

12.3.6 XRD STUDIES

X-ray diffraction patterns of Chitosan, starch, chitosan-starch beads and glutaraldehyde incorporated chitosan-starch blend film were analyzed using an X-ray diffractometer in the angular range of 10–50° (2θ) with Nickel-filtered Cu- Kα radiation(($\lambda = 1.54$nm) at voltage 40Kv and current of 30 mA (Akkaramongkolporn et al., 2000)

12.4 RESULT AND DISCUSSION

12.4.1 HARDNESS OF BEADS

The concentration of chitosan increases in chitosan/starch beads from 0.5/0.5 to 0.9/0.1. The hardness of the cross-linked beads increases as following 17.61–24.30 kg in the Table 12.2.

12.4.2 PERCENTAGE OF SWELLING

It is clear from both the Figs. 12.2 and 12.3 that percentage swelling was increased with increasing swelling time. Not only this, we observed that the percentage swelling also increases with increase in the concentration of chitosan. The protonation of -NH$_2$ group thus ensures chain penetration, leading to faster intra-hydrogen bond dissociation and efficient solvent

TABLE 12.2   Hardness of Chitosan-Starch Beads of Cross-Linked With SHMP and GA

S. No	Chitosan (%)	Starch (%)	SHMP (12.5%) mL	Glutaraldehyde (25%) mL	Hardness of beads (kg)
1	50	50	20	20	17.61
2	60	40	20	20	19.82
3	70	30	20	20	21.51
4	80	20	20	20	22.74
5	90	10	20	20	24.30

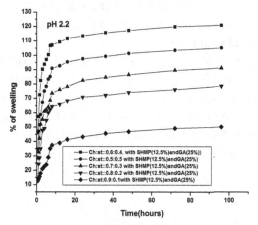

FIGURE 12.2   Percentage of swelling of beads cross-linked with sodium hexameta phosphate (12.5%) and glutaraldehyde (25%) at pH 2.2.

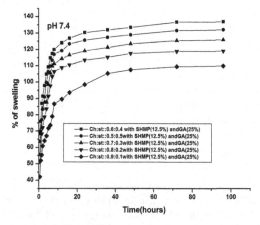

FIGURE 12.3   Percentage of swelling of beads cross-linked with sodium hexameta phosphate (12.5%) and glutaraldehyde (25%) at pH 7.4.

diffusion. In alkaline medium the swelling is mainly driven by solvent diffusion, but chain penetration due to protonation of amino group is absent. The percentage of swelling is increasing with increasing of concentration of chitosan. The percentage of swelling is more in basic medium than acidic medium.

12.4.3 FTIR SPECTROSCOPY

Figure 12.4 shows the FTIR spectra of chitosan-starch, cross-linked with SHMP and cross-linked with SHMP and GA.Spectra-1 shows chitosan-starch cross-linked with GA (25%).

The peak is obtained at 1676.1 cm^{-1} corresponding to C=O stretching of amide group. The appearance of the other peaks in chitosan depicts characteristic absorption bands at 3415.28, 2867 cm^{-1}, which represent the –OH and –CH$_2$ groups. A small peak is observed at 1413.93 cm^{-1} corresponds to CH$_3$ symmetrical deformation mode. The peak at 1151.1 cm^{-1} indicates saccharide structure and the band at 1082.3 cm^{-1} is due to C-O stretching

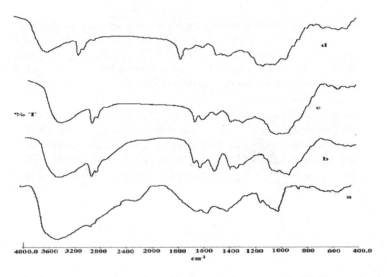

FIGURE 12.4 FT-IR spectra of (a) Beads of chitosan: starch (60:40) (b) Beads of chitosan: starch (60:40) with Glutaraldehyde (c) Beads of chitosan: starch (60:40) with and sodium hexameta phosphate (12.5%) and glutaraldehyde (25%) (d) Beads of chitosan: starch (60:40) sodium hexameta phosphate (12.5%) and glutaraldehyde (25%) with drug loaded CPM.

vibration. The peak is obtained at 1719.13 cm^{-1} corresponding to CHO stretching. Spectra-2 represents chitosan-starch cross-linked with GA (25%) and sodium hexa meta phosphate. The peak of 3399 cm^{-1} in beads indicates hydroxyl stretching vibration and the peak P-O at 1078 cm^{-1}. The appearance of other peaks at 2873 cm^{-1} and 1374 cm^{-1} indicate P=O stretch bond. Other peak at 2946 cm^{-1} correspond to –CH stretching while the peak at 1718.16 cm^{-1} is due to CHO stretching. The peak at 994.9 cm^{-1} represents the presence of an ether group in the starch. Spectra-3: FTIR spectroscopy was used to determine the interactions between starch and chitosan cross-linked beads with CPM drugs. The peak corresponding is shifted in pure chitosan 3411.81 cm^{-1} for chitosan and starch blend. The peak at 1721.31 cm^{-1} is the presence of –CHO group in Chitosan-starch. Other peaks are observed at 1569.68, 1459.51 and1364.93 cm^{-1} is due to C=C stretching, C-H stretching and C-H bending. The peaks are obtained at 897 and 761 cm^{-1} due to C-C and C-Cl stretching.

12.4.4   XRD STUDIES

Figure 12.5 shows XRD pattern of the uncross-linked and cross-linked beads.

The intermolecular interactions between the NH_3^+ and OH in chitosan and starch have limited the molecular movement of chitosan and starch chains and reduced its crystallinity chitosan-starch beads are semicrystaline in nature. Crystalinity of beads was increased continuously from chitosan-starch to cross-linked both sodium hexa meta phosphate and glutaraldehyde.

12.4.5   SEM ANALYSIS OF BEADS

The size of cross-linked dry bead is measured by SEM is 1 mm in Fig. 12.6(a). The changes in the morphology of beads are represented as a, b and c in the SEM images of Fig. 12.6(b). Chitosan/starch (0.6/0.4) of beads is as a. Samples of b and c are denoted of cross-linked beads with SHMP and with SHMP and GA both. SEM images a represents the unarranged rods. The SEM image b represents distorted hexagon network structure and also image c represents compact distorted hexagon structure.

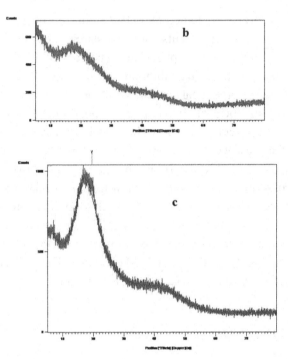

FIGURE 12.5   XRD of beads cross-linked with SHMP (a) and with SHMP and GA (b).\

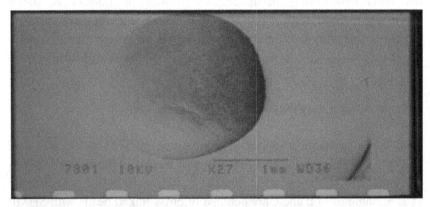

FIGURE 12.6   (a) Size analysis of cross-linked dry bead by SEM. (b) SEM (i) Beads of chitosan: starch (60:40) (ii) Beads of chitosan: starch (60:40) with Glutaraldehyde (iii) Beads of chitosan: starch (60:40) with and sodium hexameta phosphate (12.5%) and glutaraldehyde (25%).

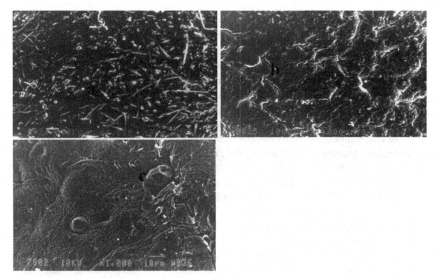

FIGURE 12.6   Continued

12.5   CONCLUSION

Chitosan-starch cross-linked beads posses a pH dependent swelling behavior. It can be successfully used for controlled drug release study. XRD and SEM results represents the as changes of crystalinity of beads by using crosslinking agents and also changes distorted network structure in SEM images. This indicates that the bead of this type is applicable in drug release study.

KEYWORDS

- beads
- chitosan
- FTIR
- starch
- swelling
- XR

REFERENCES

1. Akkaramongkolporn, P., Yonemochi, E., Terada, K., Molecular State of Chlorpheniramine in Resinates. *Chemical Pharmaceutical Bulletin*, 48(2), 231–234 (2000).
2. Atyabi, F., Manoochehri, S., Moghadam, S. H., Dinarvand, R., Cross-linked starch microspheres: Effect of cross-linking condition on the microsphere characteristics. *Arch. Pharm. Res*, 29(12), 1179–1178 (2006).
3. Barreiro-Iglesias, R., Coronilla, R., Concheiro, A., Alvarez-Lorenzo, C., Preparation of chitosan beads by simultaneous cross-linking/insolubilisation in basic pH: Rheological optimization and drug loading/release behavior European *Journal of Pharmaceutical Sciences*, 24(1), 77–84 (2005).
4. Dutta, P. K., Dutta, J., Tripathi, V. S. Chitin and Chitosan: Chemistry, Properties and applications. *Journal of Scientific and Industrial Research*, 63, 20–3 (2004).
5. Elviraa, C., Manoa, J. F., Roman, J. S., Reis, R. L., Starch-based biodegradable hydrogels with potential biomedical applications as drug delivery systems. *Biomaterials*, 23, 1955–1966 (2002).
6. Freiberg S., X. X. Zhu. A novel approach tripolyphosphate/chitosan complex beads for Controlled release drug delivery. *International Journal of Pharmaceutics*, 201, 51–58 (2004).
7. Gupta, K. C., Kumar, M. N. V. Ravi. Drug release behavior of beads and microgranules of chitosan, *Biomaterials*, 21, 1115–1119 (2000).
8. Interpenetrating Polymeric Network (IPN) of Chitosan-Amino Acid Beads *Journal of Biomaterials and Nanobiotechnology*, 2, 71–84 (2011).
9. Jameela, S. R., Jayakrishnan, A. Glutaraldehyde cross-linked chitosan microspheres as a long acting biodegradable drug delivery vehicle: studies on the in vivo release of mitoxantrone and in viva degradation of microspheres in rat muscle. *Biomaterials*, 16, 769–775 (1995).
10. Pasparakis, G., Bouropoulos, N., Swelling studies and in vitro release of verapamil from calcium alginate and calcium alginate–chitosan beads. *International Journal of Pharmaceutics*. 323(1–2), 34–42 (2006).
11. Peppas, N. A. and Colombo, P., 1997. Analysis of drug release behavior from swellable polymer carriers using the dimensionality index, *Journal of Controlled Release*, 45(1), 35–40 (1997).
12. Rani, M., Agarwal, A., Maharana, T., Negi, Y. S. A comparative study for interpenetrating polymeric network (IPN) of chitosan-amino acid beads for controlled drug release, *African Journal of Pharmacy and Pharmacology*, 4(2), 035–054 (2010).
13. Rani, M., Agarwal, A., Negi, Y. S., Characterization and Biodegradation Studies for.
14. Rinaudo, M. Chitin and Chitosan: Properties and applications. *Progress in Polymer Science*, 31(7), 603–632 (2006).
15. Rinaudo, M., Pavlov, G., Desbrieres, J., Influence of acetic acid concentration on the solubilization of chitosan. *Polymer*, 40, 7029–7032 (1999).
16. Shu, X. Z. and Zhu, K. J. Controlled drug release properties of ionically cross-linked chitosan beads: the influence of anion structure. *International Journal of Pharmaceutics*; 233, 217–225 (2002).

17. Srinatha, A., Pandit, J. K., Singh, S., Ionic Cross-linked Chitosan Beads for Extended Release of Ciprofloxacin: In vitro Characterization. Indian J Pharm Sci, 70(1), 16–21 (2008).

18. Vlachou, M., Naseef, H., Eeentakis, M., Tarantili, P. A. and Andreopoulos, A. G. Swelling Properties of Various Polymers Used in Controlled Release Systems, *J Biomater Appl.* 16, 125–138 (2001).

19. Zhang J-F., Sun, X. Z.: Mechanical properties of PLA/starch composites compatibilized by maleic anhydride. *Biomacromolecules*, 5, 1446–1451 (2004).

CHAPTER 13

DENDRIMER POLYMER BRUSHES

WEI CUI,[1] HOLGER MERLITZ,[1,2] and CHEN-XU WU[1]

[1]Department of Physics and ITPA, Xiamen University, 361005 Xiamen, China

[2]Leibniz-Institute of Polymer Research, 01069 Dresden, Germany

E-mail: cuiwei1008@163.com

CONTENTS

Keywords 225
References 226

Polymer brushes are nano-scaled surface layers created by polymers that are end-grafted onto a substrate (DeGennes, 1980). At high grafting densities, excluded volume effects make these polymers stretch away from the substrate to create a flexible and durable surface coating. While such brushes, made of linear chains, are rather well understood, the properties of brushes made of polymers with more complex architectures are still to be discovered. Good candidates for highly optimized surface layers are dendrimer brushes, of which the star-like polymer brushes are the simplest (Merlitz et al., 2011).

Recent research, based on self-consistent field theory (SCF) by Polotsky et al. (2010) delivered the result that the molecules inside a starlike polymer brush segregate into two populations: While a certain fraction of stars are stretching up toward the surface of the brush (the 'up'-population), the remaining molecules occupy the lower regions above the substrate and hardly exhibit any molecular tension (the 'down'-population). Molecular dynamics (MD) simulations by Merlitz et al. (2011), delivered details about the characteristic conformational features of these populations. Further on, it was found that the two populations are living in a dynamic equilibrium: Individual molecules are continuously flipping up or down, that is, they are changing between a fully stretched and a relaxed conformational state.

Another peculiar feature of star-like (and more general dendrimer) brushes are certain discontinuities, which show up in their monomer density profiles, if the grafting density is high and the molecules are mono-disperse. Molecules of the up-population are then highly stretched so that the grafted arms (also called 'stems') have their branching points almost located at the same vertical position. The branching then generates a local discontinuity of the monomer density that is visible as a small kink in the density profile. A gradient in the monomer density, however, results in a local osmotic pressure gradient, because each particle located inside the brush is exhibited to the thermal motions of the monomers and hence being pushed against the gradient of monomer density.

This fact is made visible with a probe particle, placed into the brush and fixed at a certain position through a stiff harmonic potential. A deviation if the particle's position from the tether point then allows to 'measure' the force acting on it. Figure 13.1 shows the effective forces acting on such

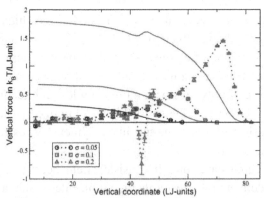

FIG. 13.1   The vertical forces (dotted curves, in $k_B T$ per LJ-length), 'measured' directly using a probe of size d=3, placed into brushes of different grafting densities σ. Solid curves are graphical integrations of the forces (shifted and re-scaled, in arbitrary units).

a probe particle of diameter d=3 (in LJ-units, the unit length being defined as the monomer diameter) inside starlike polymer brushes of different grafting densities (the grafting density denotes the number of grafted molecules per unit area). The vertical coordinates are the average positions, being close to the pre-defined tether points of the particle. Positive values of the force are pushing the particle out, toward the surface of the brush, while negative forces are pulling it back toward the substrate. The forces are reaching their maxima near the surface of the brush where the monomer density is dropping rapidly. The numerical integration of the force profile corresponds to the free energy penalty to be payed when slowly pushing the probe particle into the brush (Merlitz et al., 2012).

Note that at moderate and high grafting densities a local minimum arises in the force profile, which corresponds to a kink in the free energy curve (green). Particles, which move freely inside the brush, are being prohibited from passing freely over this area, since they have to overcome a certain free energy barrier. The analysis of how this energy barrier affects particles or molecules that are diffusing inside the brush is of great interest, because it has influence on the dynamics of molecules flipping up and down, as well as on the potential features of switchable polymer brushes which serve as environmental responsive layers.

A detailed analysis of the effects of osmotic pressure gradients is possible with chain expulsion simulations (Merlitz et al., 2008; Wittmer et al., 1994). Here, a selected individual molecule is ungrafted, that is, cut loose from the substrate, and its diffusive motion out of the brush is monitored. Such a free molecule serves as a soft probe: its average speed contains information about the overall osmotic pressure gradient, but even local osmotic effects can show up in terms of conformational changes of the diffusing probe. Naturally, since a star-like polymer brush is made of two molecular populations, it is to be expected that the diffusive motions of molecules belonging to these populations are expected to differ quite a bit.

Figure 13.2 shows the radius of gyration of released molecules (averaged over 40 independent runs) as a function of the vertical coordinate of their center of mass, and at high grafting density of $\sigma = 0.3$. The radius of gyration is split into its vertical (z) component and its averaged lateral (xy) components according to

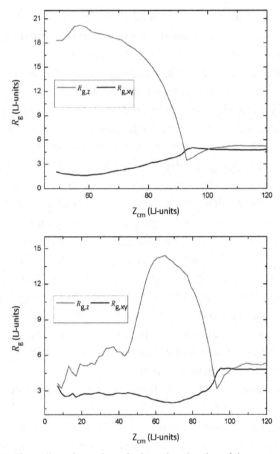

FIGURE 13.2 The radius of gyration of released molecules of the up-population (upper panel) and down-population (lower panel) on their way out of the brush, as a function of their center of mass.

$$R_{g,z} = \sqrt{\frac{1}{k}\sum_{i=1}^{k}(Z_i - Z_{cm})^2},$$

and

$$R_{g,xy} = \sqrt{\frac{1}{2k}\sum_{i=1}^{k}[(X_i - X_{cm})^2 + (Y_i - Y_{cm})^2]}.$$

The upper panel shows that initially (i.e., at lower center of mass coordinates when the molecule is just released), the vertical radius of gyration is far above its lateral component, because the molecule belongs to the up-population and is vertically stretched. On its way out of the brush, the vertical size of the star is gradually approaching its lateral size, and eventually the coil approaches isotropy, after it has left the brush (the brush surface is located near $z = 90$).

The process differs with the down-molecules (lower panel): here, vertical and lateral sizes are similar inside the lower regions of the brush, and the star becomes stretched up while it runs through the osmotic pressure gradient of the brush. Note the existence of a small vertical bump near $z{\sim}38$ LJ-units (red curve in the lower panel): Here, one part of the molecule is stretched up while it passes over the free energy barrier. At this point, the center of mass of the molecule is still located below that barrier (at roughly $z = 46$).

In recent years, the self-consistent field theory (Skvortsov et al., 1988; Milner et al., 1988; Zhulina et al., 1989; 1991; Klushin et al., 1991; Rud et al., 2013) and MD calculations (Merlitz et al., 2011; Gergidis et al., 2012) have been extended to study higher generation dendrimer brushes. Here, we restrict our analysis to MD simulations of brushes made of dendrimes with the functionality $q = 3$, spacer length $n = 50$ (monomers), and number of generatinos $g = 2$.

Be $\rho(z)$ the vertical monomer density profile, then the average layer thickness is evaluated as

$$< H > = \frac{\int_0^\infty z \rho(z) dz}{\int_0^\infty \rho(z) dz},$$

and the average monomer concentration as

$$< c > = \frac{\int_0^\infty \rho^2(z) dz}{\int_0^\infty \rho(z) dz}.$$

These results can be compared with predictions from scaling theory. In dendrimer polymer brushes, the traditional Alexander-de-Gennes scaling should apply, which in good solvent predicts layer thickness and monomer density to scale with the grafting density as $<H>{\sim}\sigma^{1/3}$, and $<c>{\sim}\sigma^{2/3}$, respectively. Figure 13.3 (upper panel) shows the monomer concentration of polymer brushes at different grafting densities. Squares are second-generation dendrimers, circles are stars (i.e., first generation dendrimers), and the two slopes are in reasonably close agreement with the scaling prediction 2/3. The same holds for the layer thickness (lower panel).

Since the scaling theory is valid, the brush density profiles can be rescaled so that they fall (approximately) onto a single (master) curve. Figure 13.4 displays the result for the second-generation dendrimer

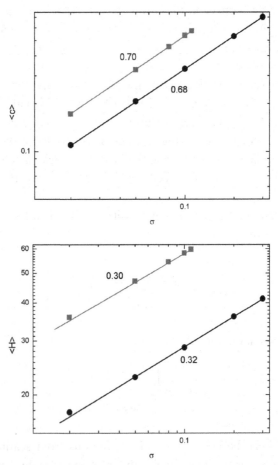

FIGURE 13.3 Monomer concentration and brush thickness of brushes, as a function of grafting density. Red squares correspond to second-generation dendrimer brushes, black circles for star-like polymer brushes

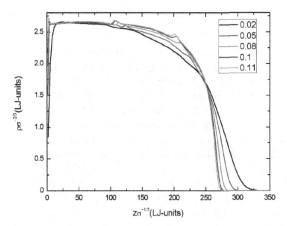

FIGURE 13.4   The rescaled monomer density profiles of second-generation dendrimer brushes at different grafting densities.

brushes. The scaling quite accurate, and it is further interesting to note that at high grafting densities not a single, but even two discontinuities of the brush profile show up: Just as is the case with star-like brushes, these kinks emerge at the branching points of the highly stretched up-population molecules, and since second generation dendrimer brushes contain two branching points, this feature is reflected twice in the density profiles. Consequently, there exist two separate free energy barriers inside these brushes.

The results presented here serve as an important step toward the functionalization of dendrimer brushes as switchable and environment-responsive surface layers.

KEYWORDS

- **dendrimers**
- **MD-simulations**
- **polymer brush**
- **self-consistent field theory**

REFERENCES

1. DeGennes, P. G. Macromolecules 13, pp. 1069 (1980).
2. Gergidis, L. N. et al., Langmuir 28, pp. 17176–17185 (2012).
3. Klushin, L. I. et al., Macromolecules 24, pp. 1549 (1991).
4. Merlitz, H. et al., Macromolecules 17, pp. 171–179 (2008).
5. Merlitz, H. et al., Macromolecules 44, pp. 7043 (2011).
6. Merlitz, H. et al., Macromolecules 45, pp. 8494 (2012).
7. Milner, S. T. et al., Macromolecules 21, pp. 2610 (1988).
8. Polotsky, A. A. et al., Macromolecules 43, pp. 9555–9566 (2010).
9. Rud, O.V. et al., Macromolecules 46, pp. 4651–4662(2013).
10. Skvortsov, A. M. et al., Polym. Sci. U.S.S.R 30, pp. 1706 (1988).
11. Wittmer, J. et al., JCP 101, pp. 4379–4390 (1994).
12. Zhulina, E. B. et al., Macromolecules 24, pp. 140 (1991).
13. Zhulina, E. B. et al., Polym. Sci. U.S.S.R 31, pp. 205 (1989).

CHAPTER 14

PHOTO-BACTERICIDAL POLYACRYLONITRILE MATRIX FOR PROTECTIVE APPARELS

G. PREMIKA,[1] K. BALASUBRAMANIAN,[1] and KISAN M. KODAM[2]

[1]*Department of Materials Engineering, Defence Institute of Advanced Technology, Ministry of Defence, Girinagar, Pune 411025, India; Email: meetkbs@gmail.com, balask@diat.ac.in*

[2]*Department of Chemistry, University of Pune, Pune 411030, India*

CONTENTS

14.1 Introduction 227
14.2 Materials and Methods 228
14.3 Results and Discussions 230
14.4 Conclusions 235
Acknowledgments 235
Keywords 236
References 236

14.1 INTRODUCTION

Surgical robes are the personal protective equipment used by both surgical patients and healthcare workers. In perioperative environment, patients are at the risk of infections by exogenous and endogenous microorganisms whereas healthcare workers are prone to occupational exposure to blood pathogens. Thus an appropriate protective surgical gown should possess effective protective barrier against transfer of microorganisms (Ramdayal and Balasubramanian, 2013). Polyacrylonitrile (PAN) a main precursor for high quality carbon fibers is well known for its intumescent property, mechanical stability, and chemical resistivity and is utilized in sophisticated textiles. Recently organic-inorganic membranes have gained immense interest as potential "next generation" membranes (Merkel et al., 2003). Inorganic antimicrobial agents like ZnO, TiO_2 have established escalating attention due to their stability under high temperature and pressure and as non-toxic agent for human and animals compared to organic substances (Roya and Montazer, 2010).

The present work describes the fabrication of bactericidal PAN using photo-bactericidal agents like ZnO, TiO_2 to demonstrate its antibacterial efficacy coupled with qualitative evaluation of molecular dynamic simulations to evaluate the interactions and the diffusion phenomenon of additives in the polymer matrix.

14.2 MATERIALS AND METHODS

14.2.1 SYNTHESIS OF BACTERICIDAL POLYMER MEMBRANES

Bactericidal membranes of PAN were synthesized by solvent casting technique. 10 wt.% PAN ($\overline{M_w}$: 150,000; Sigma Aldrich) in DMSO (Sigma Aldrich) was stirred at room temperature for 4 hr. Varying concentration of ZnO (Mw: 81.37; Merck) (100 µg/mL, 200µg/mL, 300 µg/mL) was added to this solution and further stirred for 2 hr. This mixture was casted in a glass petridish and dried at room temperature to obtain flat membranes. The resulting membranes were rinsed with DI water and dried under atmosphere conditions for 12 hr. and stored under vacuum. PAN/ TiO_2 (Titanium(IV) dioxide anatase powder, −325 mesh, Sigma Aldrich) composite membranes were also fabrication under identical conditions using the same procedure Balasubramanian (2013).

14.2.2 CHARACTERIZATION OF THE POLYACRYLONITRILE MEMBRANES

In-vitro antibacterial assay was performed against clinical isolates of Gram positive (*Bacillus subtillis* (MTCC-441)) and Gram negative (*Klebsiella pneumoniae* (NCIM-5432)) bacteria by disk diffusion technique. The microorganism's culture mediums were prepared using Muller Hinton Broth. The cultured strains were adjusted to 0.5 OD_{600}. The disk shaped samples with photo-bactericidal agents were activated under UV for 15 min and incubated in the medium for 24 hr. at 37°C. The zone of inhibition was noted after incubation period. In-vitro release test was conducted to investigate the potential of composite PAN membrane as a carrier for drug: 0.1 g of PAN membranes containing TiO_2/ZnO was added to 50 mL of phosphate buffer solution (PBS, pH = 7.4) release medium. The aliquots of release medium were spectroscopically monitored periodically using UV-Vis spectrophotometer. Release medium was replaced with freshly prepared PBS solution for continual release studies to maintain constant volume throughout the in-vitro release study (Yadav and Kandasubramanian, 2013).

14.2.3 MODELING STUDIES

Models that have been developed to explain the diffusion in polymers falls under two categories (1) molecular models that analyze the specific penetrant and polymer chain mobility (2) Free volume models that relate the diffusion coefficient with the free volume of the polymer, a phenomenological approach. Free volume plays an important role in the transport behavior of penetrant molecule in the membrane. Positron annihilation lifetime spectroscopy (PALS) is a prevalent experimental method to determine the free volume. However, the experimental technique is time consuming and cannot clearly offer the details about morphology of free volume. Molecular dynamic simulations were carried out using amorphous cell and for cite module of Materials Studio developed by Accelrys Software Inc. The COMPASS (Condensed phase optimization molecular potentials for atomistic simulation studies module) was used in the study

(Pan et al., 2008). The interaction energies between the polymer and additive and mean square displacement of the polymer chains, diffusion coefficient and free volume calculation have been performed.

14.3   RESULTS AND DISCUSSIONS

14.3.1   IN-VITRO STUDIES

Bactericidal efficacy of the PAN membranes was evaluated by in-vitro antibacterial assay (Fig. 14.1).

Pristine PAN membrane and PAN/100µg/mL TiO$_2$ demonstrated no zone of inhibition. However, composites with concentration 200 and 300 µg/mL TiO$_2$ in PAN membrane exhibited 0.8–1.0 cm zone of inhibition against both Gram positive and negative bacterial strains. PAN/ZnO (100 µg/mL, 200 µg/mL and 300 µg/mL) composites demonstrated 1.0–1.2 cm zone of inhibition against both bacterial strains. Both the photo bactericidal agents possessing well developed surface chemistry and chemical stability interacts easily with protein elements of the bacterial outer membrane and can lead to structural changes, degradation and finally cell death. Fatal damage to the microorganisms is due to highly reactive species like -OH$^-$, H$_2$O$_2$, O$_2^{2-}$ formed during exposure of composite membranes to UV light is primarily due to the formation of electron hole pair resulting from the crystal defects of ZnO and TiO$_2$. The hydroxyl radicals and superoxide are negatively charged; hence cannot penetrate the cell membrane and remain in close proximity with the outer cell membrane. H$_2$O$_2$ on the other hand can easily penetrate the cell membrane causing bacterial death.

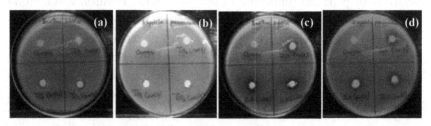

FIGURE 14.1   Digital images of Antibacterial Assay against *Bacillus subtillis* (a) (PAN/TiO$_2$), (c) (PAN/ZnO) and *Klebshiella pneumoniae* (b) (PAN/TiO$_2$), (d) (PAN/ZnO).

$$ZnO / TiO_2 \xrightarrow{h\nu} h^+ + e^-$$

$$h^+ + H_2O \longrightarrow {}^{\cdot}OH + \overset{+}{H}$$

$$e^- + O_2 \longrightarrow {}^{\cdot}O_2^-$$

$${}^{\cdot}O_2^- + H^+ \longrightarrow HO^{\cdot}_2$$

$$HO^{\cdot}_2 + H^+ + e^- \longrightarrow H_2O_2$$

Antibacterial efficacy is attributed to the leaching of the antibacterial agent from membrane and its interaction with bacterial cells. Release of ZnO/TiO_2 exhibits a lag phase with faint release followed by steady release resulting in 25% cumulative release in 24 h (as shown in Fig. 14.2). The release of bactericidal agent is chiefly controlled by the diffusion mechanism and the interaction between the additives and matrix.

14.3.2 INTERACTION BETWEEN PAN/ZNO AND PAN/TIO$_2$

When inorganic particles of micro to nano-dimensions are dispersed in the membrane, interfacial interaction between particles and polymer dictates the properties of the hybrid membrane (Desai et al., 2005). The interaction energy ΔE between PAN and ZnO/TiO_2 can be calculated as follows

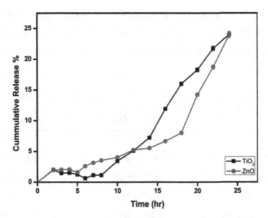

FIGURE 14.2   In-vitro release studies of PAN/TiO$_2$ and PAN/ZnO.

$$\Delta E = E_{PAN\text{-}ZnO/TiO2} - (E_{PAN} + E_{ZnO/TiO2}) \tag{1}$$

where $E_{PAN\text{-}ZnO/TiO2}$ is the potential energy of PAN/ZnO and PAN/TiO$_2$ system, E_{PAN} is the potential energy of optimized PAN and E_{ZnO}/TiO_2 is the potential energy of optimized ZnO/TiO$_2$. The particles where obtained by cutting the structural model supplied by the software. The calculation of the potential energy is based on the COMPASS force fields. The potential energy E_{Total} consists of three parts

$$E_{Total} = E_{non\text{-}bond} + E_{valence} + E_{cross\text{-}term} \tag{2}$$

where $E_{non\text{-}bond}$, $E_{valence}$, $E_{cross\text{-}term}$ are non-bond, valence and cross-term interactions respectively. Interactions between PAN and ZnO/TiO$_2$ are calculated by the Eq. (1) and the results are listed in the Table 14.1. In both cases PAN/ZnO and PAN/TiO$_2$ van der Waals, electrostatic and hydrogen bond interactions exist. From the values of energies it can be concluded that interaction energy is much higher in PAN/TiO$_2$ compared to PAN/ZnO system and the van der Waals interactions is highest in case of PAN/TiO$_2$ compared to PAN/ZnO. However, the PAN/ZnO system exhibits more hydrogen bond interactions compared to PAN/TiO$_2$ system. The cross-term interactions could be the reason for lower concentration of ZnO release compared to TiO$_2$ after lag phase.

14.3.3 MOBILITY OF THE POLYMER CHAINS

Chain mobility, which controls the diffusion paths constitutes a chief role in the transport properties of the antibacterial agents from the matrix during the release (Nagel et al., 2000). Polymer chain mobility is investigated

TABLE 14.1 Interaction Energies and Diffusion Coefficients of PAN/ZnO and PAN/TiO$_2$

	PAN – ZnO	PAN – TiO$_2$
ΔE_{Total} (kcal/mol)	−3781.004	−61,990.605
$\Delta E_{non\text{-}bond}$ (kcal/mol)	−3469.565	−61,845.727
$\Delta E_{valence}$ (kcal/mol)	−173.996	−133.596
$\Delta E_{cross\ term}$ (kcal/mol)	−137.124	−11.282
Diffusion Coefficient (m^2/s)	0.40×10^{-9}	3.26×10^{-9}

by examining the mean square displacement (MSD) of the chains as the function of the simulation time.

$$\Delta r\,(t)^2 = (r_i(t) - r_i(0))^2 \tag{3}$$

where $r_i(t)$ and $r_i(0)$ are the position of atom i at time t and 0, respectively. The bracket denotes the ensemble average, which is obtained from averaging over all atoms and all time origins $t = 0$. In the considered system, PAN is semi-crystalline polymer in which the mobility of the chains is constrained by the crystalline regions[6]. When the ZnO/TiO_2 particles are incorporated into the polymer, the chain packing perturbation plays a central role. The size of these perturbation regions corresponds to the particle size of several micro to nanometers dimensions. These regions act to disrupt the percolation of domains present in the polymer. Thus, the chain mobility increases as the crystalline region loses their percolated nature.

Figure 14.3 shows the MSDs of polymer chains, which reveal the polymer chain mobility in the hybrid membrane. The enhancement of chain mobility corresponds to the increase in disruption of crystalline regions.

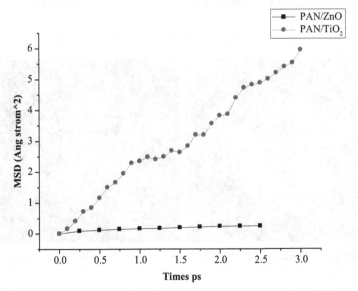

FIGURE 14.3   The MSDs of polymer chains in PAN containing ZnO and TiO_2.

Increase in chain mobility is observed in both PAN/ZnO and PAN/TiO$_2$ system. However, the disruption of crystalline region is faint in case of ZnO, whereas significant disruption occurs in PAN/TiO$_2$ system. The diffusion coefficients of ZnO or TiO$_2$ in the membranes can be calculated from the slope of the MSDs for a long time by Eq. (4).

$$D = \frac{1}{6}\lim_t \to \infty \frac{d}{dt}\sum_{i=1}^{Na} < (r_i(t) - r_i(0))^2 > \qquad (4)$$

where D represents diffusion coefficient. The diffusion coefficient of TiO$_2$ is greater than ZnO through PAN matrix (as shown in Table 14.1).

During the simulation, penetrant molecules (ZnO and TiO$_2$) were inserted into the membrane models. The diffusion runs were performed under the NVT conditions for 2 ns. The diffusion coefficient was an averaged value from all penetrant molecules (Pan et al., 2008).

Free volume plays a crucial role in the transport behavior of the penetrant molecules in the membrane. The free volume is defined as the volume on the side of Connolly Surface without atoms. The free volume morphology is shown in Fig. 14.4.

The accessible free volume in case of PAN/TiO$_2$ system is large compared to PAN/ZnO system, which signifies better diffusion transport

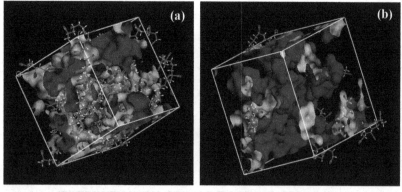

FIGURE 14.4   Simulated morphologies of free volumes (Shown in blue) in (a) PAN/ZnO and (b) PAN/TiO$_2$.

property of penetrants through PAN/TiO$_2$ membranes than in PAN/ZnO membrane. Thus increased diffusion coefficient can be attributed to higher chain mobility and larger free volume. Hence, the modeling studies predict the major parameters for diffusion transport of ZnO/TiO$_2$ through PAN and envisage that the diffusion of TiO$_2$ could be greater than that of ZnO. This is prediction is supported by experimental in-vitro release study which shows greater release of TiO$_2$ compared to ZnO after initial lag phase.

14.4   CONCLUSIONS

PAN matrix with varying concentration of ZnO/TiO$_2$ exhibited high antibacterial efficacy. Formation of hydrogen peroxide and hydroxyl, superoxide radicals on exposure to UV light implants antibacterial properties to the membrane, which exhibits potential drug carrier system by means of diffusion. The high diffusion coefficient, free volume of PAN/TiO$_2$ compared to PAN/ZnO system can result in greater diffusion transport properties of TiO$_2$ compared to ZnO. This observation in modeling studies is also supported by experimental in-vitro release study. It should be mentioned that due to the limitation of short calculation time and assumptions of polymer structures, it was rather difficult to obtain the exact value of the diffusion coefficients. Therefore, this study was simply orientated to offer the semi-qualitative and semi-quantitative trends of diffusivity in PAN matrix. Such stable PAN membranes with potential antimicrobial properties find potential application as protective apparels.

ACKNOWLEDGMENTS

Authors thank Dr. Prahlada, Vice Chancellor DIAT (DU) for his support and gratefully acknowledge DIAT-DRDO programme on Nanomaterials (EPIPR/ER/100 3883/M/01/908/2012/D (R&D)/1416 Dated: 28.03.2012) for financial assistance.

KEYWORDS

- antibacterial
- polyacrylonitrile
- titanium dioxide
- zinc oxide

REFERENCES

1. Balasubramanian K. Polymer Membrane and Process for preparing the same. Indian Patent [601/MUM/2013].
2. Desai T., Keblinski P., Kumar S. K. Molecular dynamics simulation of polymer transport in nanocomposites, J. Chem. Phys. 122, 134910–134918 (2005).
3. Merkel T. C., He Z., Pinnau I., Freeman B. D., Meakin P., Hill A.J. Effect of nanoparticles on gas sorption and transport in poly(1-trimethylsilyl-1-propyne), Macromolecules 36, 6844–6855 (2003).
4. Nagel C., Schmidtke E., Gunther-Schade K., Hofmann D., Fritsch D., Strunskus T. Free volume distributions in glassy polymer membranes: comparison between molecular modeling and experiments, Macromolecules 33, 2242–2248 (2000).
5. Pan F., Peng F., Lu L., Wang J., Jiang Z. Molecular simulation on penetrants diffusion at the interface region of organic–inorganic hybrid membranes, Chem. Eng. Sci. 63, 1072–1080 (2008).
6. Ramdayal Y., Balasubramanian K. Advancement in Textile Technology for Defence Application. Defence Sci. J. 63, 331–339 (2013).
7. Roya D., Montazer M. A review on the application of inorganic nano-structured materials in the modification of textiles: Focus on anti-microbial properties, Colloids Surf. B. 79, 5–18 (2010).
8. Yadav R., Kandasubramanian B. Egg albumin PVA hybrid membranes for antibacterial application. Mater. Lett. 110, 130–133 (2013).

INDEX

A

Acetone, 156
Acetonitrile, 81, 84, 90, 91, 94
Acrylamide, 7, 9, 88, 156, 158, 162, 165
Acrylic acid, 9, 10, 88
Acrylonitrile, 88
Acute wounds, 2
Adhesivity, 7
Adsorption process, 144, 147, 148, 149
Adsorption–desorption phenomena, 78
Aesthetics, 63, 123
Agriculture, 2, 122
Alexander-de-Gennes scaling, 223
Alginate fibers, 1, 12, 14, 144–146, 151, 152, 206, 207
Alginate in wound dressing, 1, 12
Alginate-gelatin polymeric matrix, 144
Alkalene-1 monomers, 62
Allylamine, 88
American Society for Testing and Materials, 62
Analysis of variance (ANOVA), 127, 132
Antibacterial, 7, 10, 228–232, 235, 236
Antimicrobial agent, 9, 10
Aromatic system, 35
Arrhenius plot, 25, 26
Arrhenius relation, 25
Atomic forced microscopy, 4, 15
Autolytic debridement, 3

B

Bacillus subtillis, 229, 230
Bactericidal membranes, 228
Band gap energy, 189
Beads, 7, 206–211, 213–215

Becke 3-Parameter Exchange-Correlation Functional (B3LYP) level, 81, 96
Benzenoid ring, 112
BET surface area, 63
Biaxial orientation process, 137, 141
Biaxial rotation, 121, 124, 125
Biaxially oriented polypropylene film, 136
    BOPP films, 141
    BOPP resins, 136–138, 142
    differential scanning calorimetry, 136
    gel permeation chromatography, 136
    nuclear magnetic resonance spectroscopy, 136
    xylene insoluble fractions, 136
    xylene soluble fractions, 136
Bioadhesive strength, 7
Bioadhesivity, 7
Biocompatibility tests, 7, 8
Biodegradable materials, 204
    Gaur gum, 204
    pectin, 204
    sodium alginate, 204
    starch, 204
Biological macromolecules, 144
    enzymes, 144
    nucleic acids, 144
    proteins, 144
    viruses, 144
Biological tissue, 3
Boltzmann's constant, 197
Borosil beaker, 156, 161
Box plot, 130

C

Cadmium, 145, 146, 151, 174, 177, 181, 184–186, 198

Cadmium sulfide (CdS) nanostructures, 174, 181

Calcium nitrate tetrahydrate, 145, 146

Calcium–alginate beads, 207

Cartilage replacement, 5

Casein, 154–156, 158–169

C-C stretching vibrations, 112

CdS nanostructure, 178, 181, 183–185, 188, 190

Central instrumentation facility, 169

Ceric ammonium nitrate (CAN), 154, 156, 162, 168

Chain relaxation process, 65

Charge transporting materials types, 32
    ambipolar organic semiconductors, 32
    n-type organic semiconductors (electron-transporting), 32
    p-type organic semiconductors (hole-transporting), 32

Charge-carriers, 32

Charge-transporting materials, 32, 33, 43, 55

Chemical bath deposition, 174, 176, 181, 197, 198

Chemical vapor deposition, 176

Chinese hamster ovary cells, 11
    elution test, 11

Chitin, 206

Chitin/chitosan, 12

Chitosan, 1, 11, 12, 14, 205–215
    wound healing applications, 11–14
        chito-oligomers, 11
        chitosan hydrogels, 11
        endothelial cell proliferation, 11
        fibroblast growth factor-2, 11
        lesion site, 11
        wound-healing process, 11

Chitosan-glutamic acid, 207

Chitosan-Starch beads, 207

Chloroform, 84, 90, 91, 94

Chronic wounds, 2, 14, 15

Cole-Cole plot, 22, 23

Combinatorial screening of, functional monomers, 87

Commission Internationale d'Eclairage, 37

COMPASS (condensed phase optimization molecular potentials for atomistic simulation studies module), 229

Complex permeability, 102, 104, 114, 118

Complex permittivity, 25, 100, 104, 108, 118

Computer simulations, 78, 81, 82, 94, 95

Contour plot, 130, 131

Cornstarch, 206

Coulomb interactions, 83

Council of Scientific and Industrial Research, 95

Crosslinking agents, 206, 215

Cycle time, 122, 123, 125, 130–133

Cyclic freezing-thawing method, 8

Cytotoxicity test, 8

D

Deionized water, 85, 145

Dendrimer, 33–37, 41, 55, 219, 220, 223–225

Dendritic pyrene derivatives, 35

Density functional theory, 78, 80, 96

Department of Science and Technology, 168

Design of experiments, 123, 132, 133

DFT method, 95

D-glucose, 206

Di ammonium hydrogen phosphate, 145, 146

Diabetes, 2

Dichloromethane, 81, 83, 84, 86, 90–92, 94

Dielectric analysis, 25

Differential scanning calorimetry, 136, 137, 142

Dimethylsulfoxide, 8, 90, 91, 94

Domain wall motion, 115

Domain wall resonance, 115

Dressings, 2–4, 6–9, 11–15

Drug delivery/tissue engineering matrix, 5

Dry beads, 208
Dry wounds, 2
Dynamic mechanical (or thermal)
    analysis, 6

E

Elastic modulus, 6
Electrochromic display devices, 18
Electroluminescence, 37, 54
Electromagnetic interference, 99, 101,
    102, 117
    EMI devices, 100, 118
    EMI shielding material, 103
    EMI suppression, 102
Electromagnetic wave, 102, 103
Electron dispersive spectroscopy, 145
Electrophotography, 32
Elemental analysis, 154, 159, 165
Elution test, 11
Empirical model, 127
Energy conversion units, 18
    batteries, 18
    fuel cells, 18
Environment and social setting, 2
Epithelial cells, 2
Epithelization, 2
Ethylene glycol dimethacrylate, 84, 179
Eudragit E, 7
Evolution, 20, 79
Exuding wounds, 2, 3

F

Ferrimagnetic nanoparticles, 103
Ferrite (NCZ), 109, 111, 116
Ferrite grains, 111
Fibroblast growth factor-2, 11
Fill factor, 194
Film formation mechanism, 183
Flocculation efficacy, 160, 161
Flocculation study, 160, 161
    Kaolin suspension, 160
    sewage wastewater, 161
Flower-like hierarchical morphology,
    193
Fluorescence quantum yields, 41

Fluorine-doped tin oxide, 182
Fourier Transform Infrared Spectrometer
    (FT-IR), 17, 20, 21, 28, 112, 154, 159,
    166, 210, 212, 213, 215
    FTIR spectroscopy, 18, 20, 21, 154,
        159, 166, 210, 212, 213
    FTIR studies, 21
Freundlich adsorption isotherm, 148
Freundlich equation, 148, 149
Fucoidan gels, 12
Full factorial design, 125
Fumed silica, 64, 65, 72, 75
Functional monomer, 80, 81, 83, 86–95

G

Gamma-radiation, 6
Gaur gum, 206
Gaussian 03 package, 81, 90
Gaussian 4.1 software, 84
Gel fraction, 5, 6
Gel permeation chromatography, 136,
    137, 142
Gelatin, 144–146, 151, 152
Gelation catalyst, 10
General Motors, 100
Glutaraldehyde, 206–214
Graft copolymer synthesis, 154–157,
    162, 168
Grafted casein, 154, 155, 161, 165–168
Grains, 111, 113
    ferrite grains (black), 111
    polymer grains (white), 111
Gram negative, 10, 229
    Escherichia coli, 10
Gram positive, 10, 229, 230
    Staphylococcus aureus, 10
Graphene hydrogels, 10
Gray-level co-occurrence matrix, 123

H

Hard–soft acid base principle, 145
Hazardous material, 144
    heavy metal, 144
    textile dye, 144
Healing process, 2, 9, 11, 15

Heavy metal, 144, 152, 168
Hemostasis, 2, 15
Hessian force matrix, 81
High density polyethylene, 63
Humidity and gas sensors, 18
Hybrid membrane, 231, 233
Hydrocolloid dressings, 14
Hydrogel films, 3
Hydrogels, 2–15
   cost-effective method, 3
Hydrogen atom, 91
Hydrogen bonding, 90, 91
Hydrophilic polyacrylic acid, 11
Hydroxyapatite, 144, 145, 152
Hysteresis loops, 108, 115, 116

I

Impedance analysis, 18, 22, 23
   bulk electrical resistance value, 23
   Cole-Cole plot, 22
   conductivity values, 23
   impedance plot, 23
In situ emulsion polymerization method, 100, 117
Inductively coupled plasma optical emission spectroscope, 146
Industrial equipment, 122
Inflammation, 2, 12
Infra-red spectroscopy, 4, 33, 145
Ink-jet printing, 34
Interaction energy, 87, 90, 91, 231, 232
Interaction plot, 128, 129
In-vitro studies, 230
Itaconic acid, 84, 86, 87, 88, 92, 95

J

Jar test procedures, 154

K

Kaolin suspension, 160
KBr pellets, 210
Keratinocyte migration, 12
Kinetic models, 147
   pseudo-first order kinetic model, 147

pseudo-second order kinetic model, 147
Klebsiella pneumoniae, 229, 230
Koop's phenomenological theory, 114
Korea Institute of Energy Technology Evaluation and Planning, 198

L

Lab scale extruder, 67, 68
Lagrangian equation, 147
Laminar air flow, 8
Langmuir adsorption isotherm, 148
Langmuir equation, 149
Langmuir isotherm model, 144, 148–151
Lead nitrate, 145
Lennard-Jones interactions, 83
Ligament and tendon repair, 5
Light harvesting, 35, 196
Light scattering detector, 138
Linear constraint solver (LINCS) algorithm, 83, 96
Linear low density polyethylene (LLDPE), 63, 64, 67, 70–75, 121, 122, 123, 125, 133
Liquor ammonia, 181
Low density polyethylene, 63
Luminescence signals, 35, 50, 55

M

Magnetization mechanisms types, 115
   domain wall motion, 115
   spin rotational magnetization, 115
Mark–Houwink–Sakurada relationship, 164
Mass spectrometry, 4
Materials Studio, 229
Matrix metalloproteases-2, 12
Maturation, 2
Maxwell-Wagner theory, 113
Mean square displacement, 230, 233
Mechanical properties, 4, 5, 8–10, 63, 64, 66, 74, 100, 122, 123
Medical devices, 122
Medicine, 2, 175
Melt flow index, 62, 67–70, 75, 123

Melt volume rate, 70
Merck, 84, 145, 156, 207, 228
Metallic species, 144
Methacrylamide, 88
Methacrylic acid, 84, 86, 87, 88, 96
Methanesulfonic acid, 18, 19, 20, 23, 28
Methicillin-resistant *Staphylococcus aureus*, 9
Methyl methacrylic acid, 88
Microcystin-LR, 78, 80–87, 89–95
Microwave assisted method, 154, 156
Microwave assisted synthesis, 156, 169
Microwave initiated synthesis, 155
Microwave radiation, 155–157, 162, 168
Microwave-hydrothermal method, 105, 118
Milli-Q system, 85
MIPS preparation, 84–86
    analytical grade, 85
    chemical and reagents, 84
    deionized water, 85
    HPLC grade, 85
    Milli-Q system, 85
Mobile charge carriers, 103
    electrons, 103
    holes, 103
Modulus analysis, 27, 28
Moist environment, 2, 9, 10
Molecular dynamics, 81, 95, 220
Molecular Environmental Science and Engineering Research, 96
Molecular imprinting technology, 77–80, 92, 95
Molecular weight distribution, 136, 137, 140, 142
Molecularly imprinted polymers, 78, 80, 96
Mould, 121–125, 130, 132
Mucoadhesion, 7
Mullikan atomic charge distribution, 84
Mulliken charge, 91
    hydrogen atom, 91
Multi attribute decision making, 122
MWD curve, 140

N

N,N-Dimethyl formamide, 18, 20, 28
N,Nmethylene bisacrylamide, 10
Nafion, 28
Nano filler, 64
Nanocomposite hydrogels, 8
Nanocomposites, 9, 64, 65, 75, 99–117, 179
NiCuZn ferrite, 106, 108, 111, 114, 115, 116, 117
Normal probability plot, 128
Nuclear magnetic resonance (NMR) spectroscopy, 4, 15, 136, 137, 145
    13C NMR analysis, 137
    NMR spectra, 138, 139
N-vinylpyrrolidone, 88

O

Optimized potentials for liquid simulations (OPLS), 81, 96
Optimum environment, 2
Organic electronics, 32, 35
Organic field-effect transistors, 32
Organic light-emitting diodes, 32
Organic photoconducting materials, 32
Organic semiconductor, 55
Oriented attachment process, 192

P

P. aeruginosa, 11
Particle Mesh Ewald (PME) summation method, 84, 96
Payne effect, 65
Pectin, 206
PEG, 179, 181, 182, 185–198
PEI, 179, 181, 182, 185, 188–198
Pharmacy, 2
Phosphate buffer solution, 229
Photo electrochemical solar cells, 18, 183
Photoluminescence, 35, 37
Photovoltaic devices, 32
Physical characterization of, wound dressings, 3–8

bioadhesive strength, 7
biocompatibility tests, 7
chemical/physical analysis, 4
elastic modulus, 6
gel fraction, 5
hardness, 5
mechanical properties, 4, 5
rheological tests, 6
water vapor transmission rate, 4
Plots, 25, 128–130, 132, 147, 149, 151, 183
  3D surface plot, 131
  box plot, 130
  contour plots, 130
  interaction plot, 129
  residual plot, 129
Polarized continuum model, 84, 96
Poly (methyl methacrylate), 65
Poly (styrene acrylonitrile), 65
Poly (vinyl pyrrolidone), 18, 28
Poly acrylamide grafted casein, 156
Poly carbonate, 66
Poly ethylene glycol, 10
Poly ethylene terephthalate, 65
Poly vinyl chloride, 66
Poly(hydroxyalkylmethacrylates)
  polyurethane-foam, 14
Poly(N-isopropyl-acrylamide) microgel, 7
Poly(vinyl alcohol), 14
Poly(vinyl pyrrolidone), 14
Polyacrylonitrile (PAN), 228, 236
Polyaniline (PANI), 19, 99, 100, 105–108, 112, 116, 117
Polyethylene, 11, 63, 101, 122, 123, 132, 179, 181
  glycol, 179, 181
  naphthalate, 65
Polyethylenimine, 179, 181
Polymer, 1–11, 13, 15, 18–28, 42, 47, 48, 60–67, 70–74, 78–104, 109, 111–116, 122–125, 136, 141, 155, 157, 164, 168, 169, 174, 177–185, 188–192, 197, 198, 205, 206, 209, 219, 221–225, 228–235
Polymer chain movement, 72

Polymer chains, 13, 71, 72, 109, 164, 184, 230, 233
Polymer degradation, 72, 125
Polymer grains, 111
Polymer hydrogel dressings, 1, 2
  autolytic debridement, 3
  biological tissue, 3
  exuding wounds, 3
  hydrogel films, 3
  see, hydrogels
Polymer hydrogels, 1, 2, 4, 15
  with antimicrobial activity, 1, 9
    life-threatening illness, 9
    methicillin-resistant
    Staphylococcus aureus, 9
    poly(acrylamide)/poly(N-(hydroxymethyl)acrylamide), 9
    vancomycin-resistant enterococci, 9
Polymer light emitting diodes, 48
Polymer nanocomposite hydrogels, 1, 8, 9, 101, 116
  cyclic freezing-thawing method, 8
  polyvinyl alcohol/clay nanocomposite hydrogels, 8
  PVA nanocomposite hydrogels, 8
Polymer-coated iron core-shell nanoparticles, 104
Polymeric drug delivery dressings, 1, 13
Polyolefines, 62
Polypropylene, 20, 100, 101, 136, 141, 142
Polystyrene, 101, 179
Polyvinyl alcohol, 6, 8, 9, 179
Polyvinyl pyrrolidone, 6, 20, 179
Porcine skin, 7
Positron annihilation lifetime spectroscopy (PALS), 229
Powder particle size, 122, 132
Pre-polymer complex, 83, 84, 90, 95
Pre-polymerization complex, 80, 91
Process parameter, 132, 133
Proliferation, 2, 8, 11, 12
Pseudo second order kinetic, 150
Pseudo-first order kinetic model, 147

Pseudo-second order equation, 151
Pseudo-second-order kinetic model, 144, 147
PVP, 17–25, 28, 179, 181, 182, 185, 188–198
Pyrene, 32, 35–42, 55
Pyrene moieties, 35

Q

Quantum effects, 175
Quinoid ring, 112

R

Raman spectroscopy, 145
Redox electrolyte, 183, 196
Relaxation mechanism, 27
Remodeling, 2
Residual plot, 129
Response surface graphs, 130
Rheo attachment, 67
Rheological tests, 6
Rheology, 60
Rheometers, 62
  in-line, 62
  on-line, 62
Ring types, 112
  benzenoid ring, 112
  quinoid ring, 112
Rotational moulding, 121–125, 132, 133
  agriculture, 122
  automobiles, 122
  industrial equipment, 122
  material handling, 122
  medical devices, 122
  road/highways, 122
  storage tanks, 122
Rotational moulding machine, 124

S

S. aureus, 11
S3 sample, 23, 25–27
  S2, S3, S4, S5 polymer electrolyte membranes, 26
Scanning electron microscope, 4, 15, 108, 166

Scanning electron microscopy, 107, 154, 160, 207, 209
  SEM pictures, 110, 111
Scar tissue, 2
Scherer equation, 108
Schiff base, 206
Selected-area diffraction, 106
Self-consistent field theory, 219, 223, 225
Semiconductor characterization system, 183
Semiconductor–electrolyte interface, 197
Series resistance, 195, 197
Shearing force, 66
Shielding effectiveness, 99, 100, 102, 108, 116, 117
Shunt resistance, 195
Silver, 9, 10
Sodium alginate, 145
Sodium dodecyl sulfate, 179
Sodium hexameta phosphate, 211–214
Sodium hydroxide, 105
Sonogashira coupling reaction, 36, 37
Sorption isotherm, 147
Speed ratio, 123, 125, 132, 133
Spin rotational magnetization, 115
Spin-coating printing, 34
Starch, 206–210, 212–215
Storage tanks, 122, 124
Successive ionic layer adsorption and reaction, 176
Sulfur, 38, 145, 177, 181, 184–186
Supramolecular polymeric hydrogels, 1, 13
Surface and contour plot, 132
Surface morphological properties, 190
Swelling, 8, 9, 10, 206, 207, 209–212, 215
Swelling ratio, 10
Swelling studies, 209

T

Tacticity distribution, 136–138, 141, 142
Talc-reinforced material, 100

Teflon PFA, 106
Temperature dependent conductivity, 24
Tensile strength, 6, 11, 63, 64
Tetrahedral stretching vibration, 111
Texture analyzer, 7
Thermoplastic polymers, 62
Three-dimentional surface plot, 131
Three-dimentional visualization, 92
   see, microcystin-LR, 92
Thin film characterization, 182
Thiourea, 181, 182, 184
Titanium dioxide, 236
Torque rheo attachment, 69
Torque rheometry, 66, 67, 72, 74, 75
Total dissolved solid, 168
Total suspended solid, 168
Transmission electron microscope, 4, 15
Transmission electron microscopy
   (TEM), 107
   TEM picture, 106
Triazine, 31, 32, 35, 43–66
Triazine (electrondeficient) moieties,
   32, 35
Triazine-core derived dendritic com-
   pounds, 43
Type A shore durometer, 5

U

Ultem polyetherimide, 106
United States Environmental Protection
   Agency, 80
Urocanic ethyl ester, 88
UV protection properties, 75
UV-Vis spectrophotometer, 229
UV-visible spectroscopy, 4

V

Vancomycin-resistant enterococci, 9
Verapamil hydrochloride, 207
Visco-elastic properties, 62, 65

W

Wastewater, 144, 154, 156, 161, 162,
   166, 168
Water vapor transmission rate, 4, 9, 15
Wistar rats, 11
World Health Organization, 80, 96
Wound care, 2
Wound dressing material, 4, 5
Wound healing process, 2, 14
Wound management, 2, 15
Wound types, 2, 14
   acute wounds, 2
   chronic wounds, 2
   dry wounds, 2
   exuding wounds, 2

X

X-ray diffraction (XRD), 4, 107, 145,
   146, 185, 210
   XRD method, 111
   XRD pattern, 108, 109, 148, 149,
   185, 186, 213
   XRD peak broadening, 106
   XRD studies, 210, 213
X-ray diffractometer, 106, 182, 210
X-ray photoelectron spectroscopy, 145,
   187
X-rays, 109
Xylene insoluble fractions, 136, 137
Xylene soluble fractions, 136– 139, 140,
   141

Z

Zero shear viscosity, 62
Ziegler Natta polypropylene, 137
Zinc oxide, 236